Assessment and Refurbishment of Steel Structures

Assessment and refurbishment of steel structures

Zoltán Agócs
Jerzy Ziółko, Josef Vičan
and Ján Brodniansky

Routledge
Taylor & Francis Group

LONDON AND NEW YORK

ISTER
SCIENCE
Ltd.

BRATISLAVA

First published 2005 by Spon Press

2 Park Square, Milton Park, Abingdon, Oxfordshire OX14 4RN
52 Vanderbilt Avenue, New York, NY 10017

Routledge is an imprint of the Taylor & Francis Group, an informa business

First issued in paperback 2019

Typeset in by Clara Design Studio

British Library Cataloguing in Publication Data
A catalogue record for this book is available from the British Library

Library of Congress Cataloging-in-Publication Data
Assessment and refurbishment of steel structures / Zoltán Agócs, (et al.)
 p. cm.
 Includes bibliographical references and index.
 ISBN 0-415-23598-7 (alk. paper)
 1. Building, Iron and steel. 2. Structural design. I. Agócs, Zoltán
 1938-
TA684.A78 2004
624.1′821–dc22   2003015109

ISBN 978-0-415-23598-3 (hbk)
ISBN 978-0-367-86352-4 (pbk)

Contents

List of figures

List of tables

Foreword

It is frequently claimed nowadays that more than 50 per cent of construction work in developed countries relates to the repair and renovation of the existing stock. The majority of our technical literature, however, still tends to concentrate on design and new build. Thus a text dealing specifically with assessment and refurbishment is something of a novelty – in the best possible sense of that word. Although many of the principles governing the production of cost-effective, robust, appropriate, and attractive steel construction remain valid, whether the task in hand is the creation of a new structure or the adaptation and upgrading of an existing project, there are subtle differences in the way these principles need to be applied. It is this area that the present text concentrates.

Because of the paucity of technical guidance on matters of assessment and refurbishment, this study is to be welcomed. It makes available a wealth of knowledge and expertise gained by the authors from close association with numerous actual projects. Since first-hand experience is so important as the main vehicle for developing an understanding of the intricacies of assessment and refurbishment, the sharing of this experience is a generous act.

The authors, who as will quickly become clear from the specific examples used in the text, come from the Czech or Slovak Republics and from Poland, have assembled a particularly useful collection of material. They present the basic tools of assessment, guidance on what to look for and how the findings should be interpreted, and, based on numerous actual case studies, ways in which improved performance may be provided via a whole host of different structural upgrading approaches. Such a collection of material is unique in my experience.

I suspect that the contents will be of particular interest to practitioners – especially those moving into assessment work. It must, of course, be borne in mind that each case study relates to a particular set of circumstances, but the authors have cleverly used the examples to draw out more general principles. Thus careful reading of a group of these studies should provide the reader with an awareness of the sorts of approach likely to come under consideration in the course of their own particular activities.

Many textbooks deal with structural steel design; none – apart from this one – cover assessment and refurbishment. Thus it should have wide international appeal and become a standard reference for all those organizations with an active interest in the repair and retrofitting of steel construction.

David A. Nethercot
Head of Department of Civil
and Environmental Engineering
Imperial College, London

Chapter 1

Introduction

Increasing economic and ecological pressure on the exploitation of natural resources nowadays significantly influences the aims of civil engineering. Therefore the need for reconstruction work and maintenance of engineering structures is now increasing considerably in comparison to the past.

Steel structures have an important role in civil engineering. In developing a strategy for the refurbishment of steel load-bearing structures, their positive characteristics should be noted: aesthetic elegance and prestige, economy (relatively low price of material with respect to durability and efficiency, short construction time and quick erection), flexibility (scope for reconstruction and reinforcement), environment (scope for further use of existing elements and component parts, recyclable constructional materials), safety (confidence in the use of homogeneous material, high visibility and easily accessible connections and joints) and technical convenience (easy combination with other materials, efficient anticorrosive and fire protection). Using the appropriate variant of a reconstruction method for a steel structure within the budgetary constraints can contribute significantly to the effectiveness of the solution.

When solving these difficult problems, the basic knowledge of materials, design elements and connections, as well as of metal (mainly steel) structures is assumed. At the same time the knowledge of principles of basic standards or standard regulations in the area of designing, realization, maintenance and assessment of steel structures is assumed, too.

In most cases it is more difficult and more complicated to obtain background information and to prepare a proposal for reconstruction than it is for a proposal for a new structure. This increases the financing and design responsibilities; despite the fact that steel structures offer constructional adaptability due to their high quality. By comparison with structures that use other materials, steel structures provide the widest range of reconstruction possibilities because of their material properties and their geometric and constructional adaptability.

1.1   Modern reconstruction

Scientific and technological development has been accompanied by a shortening of the moral and physical lifetime of building structures. Despite this, the design and construction of new works often neglect this fact. In the process of designing new structures, insufficient attention is paid to problems of maintenance and the possibilities for repair, reconstruction or scraping the structure or its parts. However, the difficulty of the technical problems in the preparation and realisation of a new structure can be even more formidable than in meeting the requirements for reconstruction or refurbishment of the structure.

It is clear that the adaptability of steel structures, their scope for modification, reinforcement or reconstruction is one of the most significant advantages of structures with steel load-bearing systems. Intensification of production processes, the modernization of technologies, and the increase in traffic are often accompanied by increases in service loads, the need for expansion and additional building, and refurbishment of the structure at minimum terms of realization of the proposed solution. Steel elements and components can also be used in reinforcement, refurbishment and maintenance, or in repairs of building structures with reinforced-concrete, masonry, or wooden load-bearing structures, even in cases where there has been failure or faults, or where their lifetime has been exhausted. In all such cases, the use of steel structures is effective and it can often significantly influence the overall performance of the constructional adaptations.

It is also appreciated that much production of the initial materials, i.e. the raw steel, is now redundant and so contributes to the creation of ecological imbalances and resource misallocation, which can be eliminated only by significantly reduced production in response to reduced demand. In this context, the importance of recycling materials available for re-use after exhausting the lifetime of building structures should also be noted. With respect to the total effectiveness of a project, including the complex evaluation of constructional costs, operating costs, maintenance, reconstruction, and scraping of structures, the re-use of steel can make an important contribution to the effectiveness of steel structures.

1.2   Basic terms

The following terms are used in assessing and reconstructing steel load-bearing structures:

Lifetime   The period during which a supporting structure is able to meet its original or modified function in such a way that it meets the criteria of serviceability and operational reliability

Moral ageing   Inability of a structure to fulfil new functions arising from innovation before its life time is exhausted

Assessment   A process (or set of methods) to establish the technical condition of steel load-bearing structures

Assessed load-bearing structure The load-bearing structure of a particular geometrical form, static system and dimensions at the time of assessment

Structural wear Wear due to normal operation

Structural fault A fault caused by abnormal effects-leakage of exhaust gases, local overloading, corrosion, impacts, inexpert interventions, natural events

Structural failure A change in a structure compared with its original condition that impairs its reliability

Structural flaw A deficiency in the structure caused by errors in design, construction or use

Original design documentation Documentation providing a background for the construction and erection of a structure

Documentation of actual construction The design, and other documentation in which all changes made during completion of the structure are stated

Adjusted documentation Documentation of actual completion of the structure including all changes

Loading capacity The ability of the structure to carry the required loading from the point of view of the ultimate and serviceability-limiting conditions on the operation of the static and dynamic loading system

Opinion on loading capability of an existing structure A judgement on an existing load-bearing structure with respect to its ultimate and serviceability limiting conditions, perhaps with respect to its lifetime

Reconstruction (refurbishment) of a steel structure A set of various proposals by which the layout, static system and loading capacity of elements and parts of an existing structure may be modified with the intention of meeting changed requirements for the usage and operation of the structure.

Refurbishment of steel structures can be divided into:
- Reinforcement
- Renovation
- Extension
- Replacement
- Relocation
- Special arrangements.

Reinforcement of the structure An adaptation intended to recover operational features of a structure or to increase the increasing loading capacity, reliability and serviceability of the structure

Direct reinforcement The reinforced and reinforcing parts that constitute one load-bearing element

Indirect reinforcement An arrangement in which the reinforced and reinforcing parts of the structure are constituted by separate load-bearing elements

Structure securing A temporary arrangement to secure the reliability of a structure

Renovation   An arrangement of structural parts or the structure as a whole after extraordinary events and accidents

Extension   (of a superstructure and additional building) An extension of an existing section of the structure

Replacement   An arrangement that replaces the original structure or a section of it by a new structure linked to an existing layout

Relocation   A part of the original or the entire original structure are moved

Special arrangements   include:

- Indirect reinforcing
- Changes to the static system
- Prestressing
- Reinstallation of the structure for other uses.

1.3   Principles of the design and assessment of steel structures for refurbishment

In designing the refurbishment of an existing structure we should follow the present valid standards and regulations. Previously valid standards or regulations can serve only as informative background. A project for reconstructing a steel structure must be documented by an examination of the reliability of the original supporting structures.

The proposal for reconstruction should be preceded by a substandard analysis of the suitability or unsuitability of the structure for reconstruction. Structures unsuitable for reconstruction, for which replacement is recommended include:

- Structures significantly damaged by impact or explosion
- Structures totally damaged or distorted by high temperatures during a fire
- Structures totally corroded
- Structures for which the condition of stresses, or the history of variable loadings cannot be reliably determined,
- Structures of iron wrought and other low-quality materials (with the exception of rare historical structures)
- Structures for which it is not possible to obtain necessary results of assessing examinations, taking into consideration the level of acceptable costs and economics of examining the existing load-bearing system.

As background in preparing a proposal for reconstruction, the structure should be examined to established:

- The condition of the structure as a whole
- Properties of the materials and subsoil
- Loads that affect the structure or have affected it in the past
- The influence of the environment on the structure
- Documentation for the structure.

The preparation of the reconstruction proposal should cover both the preliminary and the comprehensive assessment.

Reconstruction proposals, particularly in industrial areas, mostly involve modernization of technical devices. When the weight of new technological devices is less than that of original ones, there will be reserve capacity and so a reserve in the load-bearing strength of the original steel load-bearing structure, which it may be possible to use in accommodating heavier production equipment or for other purposes.

Given these considerations, a viable proposal for interventive reconstruction of an existing structure must observe not only valid standards for the loading on both building and bridge structures, but also precise data from the client on the required arrangement of technical devices, their weights and the loading processes characterizing the final user's regime for the structure.

Verification of the influence of actual operating conditions and loading effects on the dimensions and functioning of the structure (as calculated for the structure) must be always carried out in the following cases:

- In proposals for reconstruction
- In proposing a repair if the arrangement of the loading of the structure has changed
- When faults and failures reduce structural reliability have occurred
- When standard loading values are changed due to the change of use of the object.

When the use a structure is changed, it is possible to reduce the required calculation: by comparing the force effects of calculated loads acting in the structure before the change and after it. In those parts of the structure where increased effects of force loadings are found, a complete calculation of verification of dimensions is required.

1.3.1 Materials used in steel structures since the 1920s

Progress in the construction of steel structures was accelerated by expansion in the industrial manufacture of metals after 1785, mainly of cast iron and wrought iron. By the end of nineteenth century steel began to be produced using various technologies, such as the crucible and processes devised by Bessemer, Siemens-Martin and Thomas. Later, after the World War II, the Linz – Donawitz (LD) system was introduced and was used in Eastern Europe until the 1980s, when it was replaced by continuous steel casting. This steel was a higher-quality material than wrought iron, especially with respect to its view mechanical properties and the stability of chemical composition. Since 1926 one standardized name – steel – has been used for all forged iron manufactured in the fluid state.

The mechanical properties of manufactured steels were gradually defined by various regulations and later by technological standards. For the design of steel structures, it was particularly important to determine the strength characteristics of manufactured steels, in order to use the method of allowable stresses for assessing reliability, the values of allowable stresses were also needed. The first document that defined these values was the Regulation published in 1904 in Vienna, which was translated and published in 1911 in Prague [14]. It contained the first overall review of allowable stresses of materials used in the construction of bridges. Further improvement of the properties of steel was recognised in the decrees issued in 1915 and 1921, and the decisions handed down in 1921 and 1923. The values for allowable stresses were graduated according to the span of bridge elements because in these values dynamical effects were also taken into consideration. Maximum values of allowable stresses were limited at absolute values of 90 MPa for wrought iron and 100 to 115 MPa for fluent iron.

A considerable change in design methods and values for allowable stresses appeared as late as 1936, with the introduction of Czechoslovak Standard 1230. A uniform regulation for bridges [15] was complemented in 1939 by regulations for the welding of bridge structures. Here, allowable stresses were defined by absolute values, depending on the load combination and type of steel. This standard was valid until 1950 when it was replaced by a regulation for railway and road bridges Regulation for Bridge Design. In 1953 this regulation was replaced again by Czechoslovak Standards 73 6202 [16] and 73 6204 [17].

In 1972, in the design of steel bridges the assessment of reliability according to the method of partial safety factors began to be used. This was incorporated in Czechoslovak Standard 73 6205 [18], the last amendment of which was in 1987 [19]. This standard is currently valid, together with the European prestandard [12] for steel bridges. In these standards the characteristic values for steel strength, as well as the corresponding partial factors for material resistance, are defined according to the method by which the design values are determined. To verify the reliability of existing bridge structures as part of their complex evaluation, using the correctly valid method of partial safety factors, the characteristic and design strength values of steels produced since the beginning of twentieth century are important. They are given in Table 1.1

The design values for steel strength from Table 1.1 can be replaced by more accurate values, the determination procedure for which is given in Section 1.3.2.5 below. The more accurate values can be used on condition that there are regular inspections at periods determined by the corresponding standards, and that the bridge structure is in trouble-free technical condition.

Table 1.1 Characteristic and design strength values for steel

Year of production	Material strength class		Basic allowable stress σ_{allow}[MPa]	Characteristic strength value f_{yk} [MPa][2]	Design strength value f_{yd} [MPa][2]
Until 1904[1]	Wrought iron		130	210	180
Until 1937	Float steel		140	230	200
1938–1950	S 235		140[3]	230	200
	S 355		195	335	265
	S 235	Thickness in mm ≤ 25	140	230	200
		> 25	130	210	180
1951–1969	S 355	≤ 16	210	360	285
		> 17	200	340	270
	S 235	≤ 25		260	210
1969–1986		> 25		250	200
	S 355	≤ 50		353	280

1   For float steel from the period after 1 January 1895 and before 1 January 1905
2   f_{yk} = 230 Mpa and f_{yd} = 200 MPa
3   Only if this value was used in static calculations

1.3.2 *Reliability of building structures*

1.3.2.1 Introduction

In general, by the term 'reliability of building structure' we understand the ability to meet the required functions while preserving their service indicators in the given conditions over the required time interval. The basic function of a building structure is to carry safely all loads affecting it according to the purpose for which it has been designed. Accordingly, the reliability of building structures is a comprehensive indicator that includes various partial components, such as safety, usability, lifetime (or durability), and economy. A building structure is considered to be reliable if for its planned lifetime it is safly, and economically designed, and is constructed to be fit for the purpose required by the project. Structures can be in various condition during their lifetime, trouble-free or in trouble. The limit state is a specific condition of a structure when the structure does not meet the proposed operational requirements any more, so it stops performing the required functions. The origin of the limit state is influenced by external and internal causes. The external causes of the limit state include the loading of the structure, influence on it and its environment. The internal causes most frequently include properties of the materials, and of the structure and human factors. As the causes of the limit state, as well as the limiting conditions of the structures themselves, are random phenomena, it is necessary to use probability theory and mathematical statistics in their examination.

1.3.2.2 Methods of assessing the reliability of structures

Although the mathematical theory of reliability has been developed to a very high level, it cannot be fully used for assessment of reliability of structures. At present there is no realistic mathematical model describing the reliability of structures as a whole. Therefore reliability is quantified by using conditions of reliability that are applied to individual elements of the structure. If these partial elements meet the corresponding conditions of reliability, the whole structure is considered to be reliable. The corresponding conditions of reliability are derived from the definition of the limit state for the structure and depend on the methods of reliability assessment. These are divided into:

- Deterministic
- Probabilistic.

The most-used deterministic method is the method of allowable stresses. The majority of existing building structures were designed according to this method, which under the prevailing conditions was used until 1970. As a deterministic method, it uses a general condition of reliability:

$$F_k < R_k / \gamma \tag{1.1}$$

where:

E_k are characteristic loading effects
R_k is the characteristic resistance of a material (for steel it is the nominal value of its yield stress)
γ is a conventional safety factor.

The conventional safety factor γ considers all uncertainties in input parameters involved in the process of checking reliability, without detailed specification of the proportion attributable to individual components. Simultaneous occurrence of various variable loads was considered by means of two combinations of loads: principal load and overall load. A greater number of variables short-term loads involved in the overall load combination load was reflected in a lower value of γ. From this description it is obvious that the method of allowable stresses resulted from empirical experience obtained in the development of the design of building structures. The effect of random variability in input parameters was not statistically evaluated because there were no standard statistical characteristics; hence it was necessary to use the empirical basis.

Gradual development in the theory of designing building structures and increasing pressure on the economics of the design proposals prompted research in the theory of reliability of structures. The qualitative shift was brought about by the application of mathematical and engineering approaches in the theory of reliability to the assessment of reliability of building structures. In general, these approaches are considered to be probabilistic methods of assessment of structure reliability.

According to the level of the theory of probability they use, they can be classified into three levels, as in Figure 1.1.

The classification of the methods of reliability in Figure 1.1 shows the mutual connections between the individual levels of the probabilistic methods, and their relation and connection to the deterministic methods. Current standard practice in checking reliability uses the methods of the first level, which uses the outputs of higher levels as well as empirical experience for using the method of allowable stresses.

The basic method of assessing the reliability of building structures is the method of partial factors determined as a method of limit states. This method has a probabilistic basis but it is used in a deterministic form. The probabilistic method of partial factors derives from engineering methods of the theory of reliability from which this method was developed.

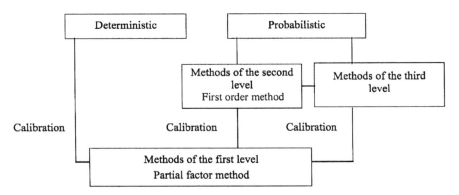

Figure 1.1 Classification of methods in the theory of reliability

1.3.2.3 Engineering methods in the theory of reliability

Engineering approaches to the theory of reliability regard all input quantities for the process of checking reliability of structures to be random variables, thus neglecting an important factor to do with structure lifetime, namely time. To determine the reliability of an element of the structure, they use the reliability margin Z in the relation

$$Z = R - E, \tag{1.2}$$

where:

R is the resistance of the element generalized as a function of its components

E is the generalized response of the element to loading as a function of the partial effects of loading.

A quantitative expression of the reliability of an observed element is given by

$$P_f = P\,(R - E < 0) \le P_{fd},$$ (1.3)

Where P_f is the probability of failure of the observed element

P_{fd} is a design value of the probability of failure defined by standards for designing building structures [1], [2].

Assuming there is a known distribution of the structural element resistance R and its load effects E, the distribution of the reliability margin Z can be determined. For the known distribution of the reliability margin Z, the probability of failure P_f of an element is determined by an expression defining the dashed area in Figure 1.2:

$$P_f = \int_{-\infty}^{0} f_z(z)\, dz,$$ (1.4)

while the probability of reliability P_r will be given by

$$P_r = 1 - P_f = 1 - \int_{-\infty}^{0} f_z(z)dz = \int_{0}^{\infty} f_z(z)dz$$ (1.5)

In Figure 1.2 another possibility is also indicated, to express quantitatively the reliability of an element by means of the reliability index according to Cornell [4], which has the form

$$\beta \ge \beta_d$$ (1.6)

where:

β is a reliability index defined by the expression

$$\beta = (\mu_R - \mu_E)/(\sigma_R^2 + \sigma_E^2)^{0,5}$$ (1.7)

β_d is a design value of the reliability index defined by standards for designing building structures

$\mu_R\,(\mu_E)$ is the mean value of resistance of the element (load effects of element)

$\sigma_R\,(\sigma_E)$ is the standard deviation of the element resistance (load effects of element).

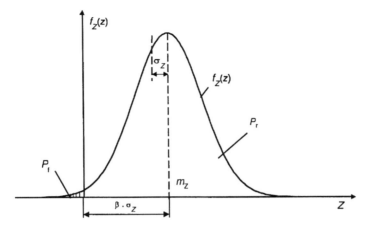

Figure 1.2 Definition of the reliability index β and probability of failure P_f

The reliability index β and probability of failure P_f are mutually dependent according to the relation

$$P_f = \phi\,(-\beta) \text{ or } \beta = \phi^{-1}(P_f), \qquad\qquad (1.8)$$

where φ is the cumulative distribution function of the probability density function

The design values for the probability of failure P_{fd} and reliability index β_d for designing steel structures with a planned design lifetime $T_d = 80$ years depend on the reliability level (the importance of the structure) and the kind of limit state in Table 1.2.

Table 1.2   Design values for the probability of failure P_{fd} and reliability index β_d for steel structures

Reliability level	Limit states			
	Ultimate		Serviceability	
	P_{fd}	β_d	P_{fd}	β_d
Decreased	50×10^{-5}	3.30	16×10^{-2}	1.00
Basic	7×10^{-5}	3.80	7×10^{-2}	1.50
Increased	0.8×10^{-5}	4.0	2.3×10^{-2}	2.00

The use of engineering methods for the theory of reliability has until recently for practical applications been limited by the complicated expression for the randomly variable reliability margin. It should be noted that the resistance R as well as the effects of loading E are functions of randomly variable partial components, so the reliability margin is in general a function of random quantities X_i. The condition of reliability is often of the form

$$Z = \prod_{i=1}^{k} X_i - \sum_{i=k+1}^{m} X_i \geq 0 \tag{1.9}$$

which is complicated by the nature of Z as well as by the probability of failure P_f, which is then expressed by the general relation

$$P_f = \int_D f(X_1, X_2 ... X_n)\, dX_1.dX_2..dX_n \tag{1.10}$$

where f $(X_1, X_2 ....X_n)$ is a function of the compound probability density function of random variables of $X_1, X_2 ...X_n$

The direct substitution of values in expressions (1.9) and (1.10) is unrealistic, so approximate solutions given by analytical or numerical methods are used. The analytical approaches use approximate methods because they are either first order reliability methods or second order reliability methods. The numerical approaches use various types of simulation techniques, such as the Monte Carlo method and its modified versions in the form of Latin hypercube sampling, importance sampling, response surfaces and others.

1.3.2.4 The partial factor method

In present conditions, the partial factor method has been used to design building structures since 1970. As it is of a probabilistic nature, but is applied in a determi-nistic form, it is ranked among probabilistic methods of the theory of reliability. The effect of randomness of quantities entering the process of reliability verifica-tion takes into consideration loads and resistance, using partial safety factors. Though the idea of using partial factors was introduced by N.S. Streleckij as early as 1947, their regular use began only in the second half of the twentieth century. Once the partial factor method was introduced into European normalization pro-cedures [1], it became the standard method for verifying the reliability of building structures and bridges.

The general condition of the reliability of a building structure is defined in the method of partial factors in a separate relation form

$$E_d \leq R_d, \tag{1.11}$$

where:

E_d are design load effects

R_d is the design resistance of the material, element or structure

This separated reliability condition makes it possible to investigate independent-ly the load effects and the resistance of the structure. Nevertheless, at the same time it represents a retreat from the fully probabilistic form of the general condition of reliability given in expression (1.2). Separation is attained through the so-called separation (sensitivised, linearised) functions of load effects and resistance of structure.

Using the separation function the separate reliability condition can be derived as

$$\mu_E + \beta_d \alpha_e \sigma_E \le \mu_R - \beta_d \alpha_R \sigma_R \qquad (1.12)$$

where:

$$\alpha_E = \frac{\sigma_E^2}{(\sigma_R^2 + \sigma_E^2)^{0.5}} \qquad \text{is a separation function of load effects}$$

$$\alpha_R = \frac{\sigma_R^2}{(\sigma_R^2 + \sigma_E^2)^{0.5}} \qquad \text{is a separation function of structural resistance.}$$

The left-hand side of expression (1.12) represents a design value for load effects E_d, while the right-hand side is a design value for structure resistance R_d. A graphical representation of condition (1.12) is shown in Figure 1.3. Separation functions α_E and α_R are replaced in the method of partial factors by constants $\alpha_E = 0.7$ and $\alpha_R = 0.8$, expressing very well the real form of the functions α_E and α_R within the range $0.16 < \sigma_E/\sigma_R < 7.6$, while applying to the range of reliability index $\beta = 3.8 \pm 0.5$.

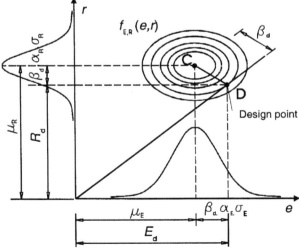

Figure 1.3 Graphical representation of the separated reliability condition

The separated reliability condition in the form (1.12) is of probabilistic nature. However, it acquires a deterministic shape via separating the load effect and element resistance. Its practical application is limited in normal approaches due to lack of knowledge of the statistical characteristics of the fundamental input parameters, especially in case of load effects. This means we are not able to set design values for those parameters corresponding to the design level of structure reliability or, as the case may be, to the design value for probabilistic failure P_{fd}. This limitation can be removed by applying partial safety factors for load and material. Once they are introduced, the procedure for setting design values of load effects and structure resistance is made more general, for they allow the use of empirical experience obtained by applying deterministic methods (especially the method of allowable stresses). Design values are then set, using values of characteristic and relevant partial factors. A characteristic value is understood here as a value corresponding to a fractal of dividing random quantity prescribed by norms, provided this distribution is known. If we do not know the distribution, the characteristic value is understood as one to be specified based on experience acquired by applying deterministic and empirical approaches to verifying reliability of building structures. Characteristic values of loads of building structures and materials resistance are specified by the relevant norms for designing building structures, [1], [2] and others. At the same time, they also define the values for appropriate partial safety factors for loads and materials. Design values for load effects and structure resistance are defined in the method of partial factors as follows:

$$E_d = \gamma_{Ed} E(\gamma_f F_k, a_d) = E(\gamma_F F_k, a_{nom}) = \gamma_F E_k \tag{1.13}$$

$$R_d = R(X_k / \gamma_m, a_d) / \gamma_{Rd} = R(X_k / \gamma_M, a_{nom}) = R_k / \gamma_M \tag{1.14}$$

and where

E_k and R_k are characteristic values of load effects and structure resistance,

γ_f is a partial safety factor for load effects considering adverse deviations of loading from their representative values,

γ_F is a partial safety factor for load effects, considering model uncertainties and dimension deviations,

γ_m is a partial factor for material properties,

γ_M is a partial safety factor for materials, considering model uncertainties and dimension deviations,

γ_{Ed} is a partial factor considering uncertainties of the model of load response,

γ_{Rd} is a partial factor considering uncertainties of the resistance model,

a_{nom} is the nominal value for a geometric characteristic,

a_d is a design value for a geometric characteristic,
         where $a_d = a_{nom} \pm \Delta a$.

On condition of the proportionality of load effects E to the action F and model uncertainties, we get the following relations for the partial factors γ_F and γ_M

$$\gamma_F = \gamma_{Ed}\gamma_f(1 + \Delta a/a_{nom}) \tag{1.15}$$

$$\gamma_M = \gamma_m\gamma_{Rd}/(1 + \Delta a/a_{nom}) \tag{1.16}$$

1.3.2.5 Application of the partial factor method in assessing the reliability of existing structures

In assessing reliability of existing structures, it is necessary to consider certain differences they may have in comparison with newly designed structures. The primary differences are:

- Existing structures are periodically checked via inspections and diagnostic inquiries whose results represent a source of information that can reduce uncertainties in the input parameters involved in verifying their reliability.
- The reliability of the existing structure is verified for residual lifetime, not for its planned lifetime, as is the case with a design of a new structure.

When verifying the reliability of a new structure, it is necessary to consider possible imperfections of the system or its elements. These are to be considered in terms of the relevant by norms for the various design values. With existing structures, this requirement is extended to consider the present technical condition of the building upon verifying its reliability. Imperfections of the system and its elements, as well as failures and damage, can be measured on a real structure and these measurements can be inserted directly into transformation models of structure response to load or the resistance of the structure and its elements.

Influence of periodic supervisory activities
Considering the results of supervisory activities provides an opportunity not only to correct transformation models of the structure for load and resistance, but also to reduce uncertainties about the geometric and material properties of the structure and its loading. Thus, there is the opportunity to verify the existing structures to a lower level of reliability than it is the case with the newly designed ones, for which this information is not known.

The significance of the information obtained from inspections and diagnostic investigations of the reliability of the existing structure can be shown using a mathematical model derived in [5] and further elaborated in [6] and [7]. The theoretical model assumes a fundamental design level of reliability defined according to Table 1.1 for a design value of probability of failure $P_{fd} = 7.0 \cdot 10^{-5}$ or for a design index of reliability $\beta_d = 3.80$. An inspection carried out at time $t_{insp} < T_d$ Where T_d is the planned lifetime of the structure, $T_d = 80$ years) has shown that the verified structural element should not fail in the sense of exceeding any of its limit states.

However, minor damage due to degradation of material, for instance corrosion of steel may occur. Given the residual lifetime of the structure, it is possible, using the above information, to specify the probability of failure by using the conditional probability after an inspection, given by

$$P_{fu}(t) = [P_f(T_d) - P_f(t_{insp})]/[1 - P_f(t_{insp})],$$ (1.17)

to which corresponds the reliability index

$$\beta_u(t) = \phi^{-1}[P_{fu}(t)]$$ (1.18)

In (1.17) and (1.18),

$P_f(T_d)$ is the probability of a failure of the verified element within planned life time $T_d = 80$ years

$P_t(t_{insp})$ is the probability of a failure of an element for the period of time $t = t_{insp}$.

Where

t_{insp} is the time at which the inspection of the structure is carried out.

Calculations of the above probabilities of failure were presented in [5], [6] and [7] for a time-variant element resistance $R(t)$ and load effects $E(t)$ being normally distributed random variables.

It was shown by the following parametric studies, of which some results are shown in Figure 1.4, that the reliability index $\beta_u(t)$ for residual lifetime increases after carrying out an inspection of structure, i.e. its reliability increases. This increase is a result of the information acquired upon the inspection.

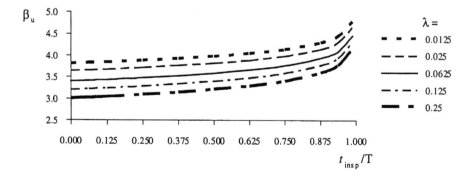

Figure 1.4 Increase in the reliability index of a structure in reaction to the time of carrying out an inspection

If proper implementation of inspections can be assumed when designing a structure, the structure can be designed to a lower target reliability index β_t. This can be determined by iteration, so that at the end of the life-time, given that inspections are carried out, its value did not decrease below the basic design level $\beta_d = 3.80$. Figure 1.5 shows the decrease in the reliability index b_t for the residual lifetime of the structure after carrying out an inspection within time t_{insp}/T_d. The parameter λ (t) in both figures represents the intensity of load effects within the requisite time and thus also the intensity of failures.

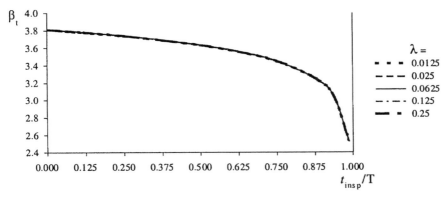

Figure 1.5 Decrease in the target reliability index for verifying the reliability of existing structures

Effect of residual lifetime of the structure
The design lifetime of building structures is an assumed period of time within which these structures are used for their designed purpose with appropriate maintenance, and, without the need for more extensive renovation. The planned lifetime of building structures is usually estimated to be in the range of 80 – 100 years, although some parts of a structure may have a shorter lifetime. This means that the design already assumes reconstruction of certain parts of it (e.g. the reinforced concrete slabs of a steel bridge deck). The residual lifetime of structure is understood as the time for evaluating it after the assessment examination to the end of its active use. So the (tentative) estimated residual lifetime is the difference between the design lifetime and the time during which the structure was in operation, provided all design requirements were respected (purpose of the structure, periodic inspections, current maintenance, etc.). The true residual lifetime is not a period of time that can be exactly determined. It is affected by many factors of an objective as well as a subjective nature. Maintenance ranks particularly important among the subjective ones affecting the residual lifetime of a structure. Administrative interventions including changes to the purpose of structures, modifications of design standards for building structures and changes in operating conditions are objective factors affecting residual lifetime.

Knowledge of the true residual lifetime is important in making a decision on the optimum time for reconstruction. However, determining it is very complicated even with present means. Various models have been devised for the time modification of the reliability of bridge elements due to degradation of material, such as by corrosion of steel or of concrete reinforcement [5], [6]. However, these models have not been verified in practice, because degradation processes of materials are long-term phenomena, so research into them takes a very long time. Attempts to predict residual lifetime remain largerly theoretical providing little scope for the development a standardised approaches.

The level of structural reliability is in all cases connected closely with a structure's lifetime. In new structures, it is their planned lifetime, while in existing ones it should be the residual lifetime for which the determined level of reliability is applicable. As it is still not possible to determine realistic residual lifetimes, instead the reliability level for planned residual lifetimes of the existing structures.

The reliability level for existing structures is dependent on the time of inspection, so it corresponds to a certain planned residual lifetime of the bridge. However, the period of evaluation varies and the level of reliability changes with the assumed residual lifetime of the bridge. Using a more detailed analysis of these dependencies, we have specified more precisely the level of reliability according to the assumed residual lifetime of the structure. The values acquired from parametric studies are shown in Table 1.3.

Table 1.3 Reliability level according to the residual lifetime of the structure

Residual lifetime (years)	Evaluation carried out after							
	The twentieth year		the fortieth year		the sixtieth year		the seventieth year	
	β_t	P_{ft}	β_t	P_{ft}	β_t	P_{ft}	β_t	P_{ft}
5	3.37	3.73 E-04	3.21	6.70 E-04	3.10	9.65 E-04	3.06	1.11 E-04
10	3.51	2.23 E-04	3.37	3.71 E-04	3.28	5.18 E-04	3.24	5.91 E-04
20	3.62	1.48 E-04	3.51	2.21 E-04	3.44	2.95 E-04	–	–
30	3.67	1.23 E-04	3.58	1.73 E-04	–	–	–	–
40	3.69	1.11 E-04	3.62	1.48 E-04	–	–	–	–
50	3.71	1.05 E-04	–	–	–	–	–	–
60	3.73	9.70 E-05	–	–	–	–	–	–

For practical use in assessing the reliability of existing structures, we recommend the following values (Table 1.4).

Table 1.4 Recommended values for the reliability level of existing structures

Residual lifetime	β_t	P_{ft}
20 years	3.50	$2.30 \cdot 10^{-4}$
10 years	3.40	$3.40 \cdot 10^{-4}$
5 years	3.20	$6.80 \cdot 10^{-4}$
less than 5 years	3.00	$1.40 \cdot 10^{-3}$

Effect of failure and damage to structures

The effect on reliability of failure and damage to structures discovered in the diagnostic examination are most appropriately shown when reliability is assessed in the true technical condition of the structure. Failures and damage to elements of the structure that is being assessed should be projected into the calculation model of the global analysis as well as into the calculation models of reliability of the affected structural elements and their cross-sections. With respect to the calculation model of the global analysis, consideration include the changes to the toughness of elements due to damage (e.g. by corrosion of steel) that affect redistribution of internal forces within the structure considered as a spatial system. The influence of failures and damage is substantially more significant in the models of resistance than in those of global analysis. Alterations in the geometric parameters of cross-sections (corrosion losses) or elements (deviations from direct alignments) considerably affect the reliability of structural elements and their cross-sections in a very distinct way. Methodologically, the procedure is the same as when considering the initial imperfections of systems and their elements in designing a new structure. The basic difference is the possibility of specifying the damage and introducing it as the appropriate imperfection in calculation models of global analysis and structure reliability.

Modification of values of partial safety factors for existing structures

Adaptations of the reliability level of an existing structures in the method of partial factors can be projected into the design values for load effects and resistance of the structure. As these are determined by means of the characteristic values and appropriate partial safety factors, this adaptation will be expressed in a change in values for the partial safety factors of load effects and resistance of the structure.

In checking the reliability of building structures, the European pre-standard [1] has introduced two values for the partial safety factors of loads for persistent and transient design situations expressed in terms of ultimate limit states. In particular

- For adverse effects of permanent loads: $\gamma_{FG} = 1.35$
- For adverse effects of variable loads: $\gamma_{GQ} = 1.50$.

These values for partial safety factors are also essential in checking the reliability of existing structures. In addition to them, the European pre-standard specifies namely states [1] other values for these factors, applicable for checking the static balance of the structure or, as the case may be, to check the reliability of its foundations, as well as the values of these factors for accidental design situations. When applying an adjusted reliability level for existing structures, the values of the above-mentioned partial safety factors of load effects will be modified. Provided the value of the coefficient of variation of permanent load effects is unaltered which is recommended in [1] as $v_G = 0.1$, the following values may be set depending on the planned residual lifetime of the structure t_r:

- For $t_r > 20$ years γ_{FG} @ 1.30
- For $10 < t_r \leq 20$ years γ_{FG} @ 1.30
- For $5 < t_r \leq 10$ years γ_{FG} @ 1.30
- For $t_r \leq 5$ years γ_{FG} @ 1.25.

Similarly, for the partial safety factors of variable load effects, given on unaltered value of the variation coefficient v_Q of variable load effects, values may be set again depending on the planned residual lifetime of the structure:

- For $t_r > 20$ years $\gamma_{FG} \cong 1.40$
- For $10 < t_r \leq 20$ years $\gamma_{FG} \cong 1.35$
- For $5 < t_r \leq 10$ years $\gamma_{FG} \cong 1.30$
- For $t_r \leq 5$ years $\gamma_{FG} \cong 1.25.$

Note: These values for the partial safety factors of permanent and variable load effects were rounded to the nearest 0.5.

It is also possible to modify the values for the partial safety factors of the material. For steel structures, their values are defined by the European pre-standard [11] or by the relevant national standards, in our case by the standard [2].

The European pre-standard gives one value $\tilde{a}_{M0} = \tilde{a}_{M1} = 1.10$ for all kinds of steel considered in this standard, i.e. S 235, S 275 and S 355, for the partial safety factor of material with respect to ultimate limit states. The standard [2] gives different values for the partial safety factors of materials, depending on the kind of steel, but similarly to the European pre-standard [11], does not differentiate among these values according to the kind of cross-section. The values recommended by the standard [2] are as follows:

$$\gamma_{M0} = \gamma_{M1} = 1.10 \qquad \text{for steel S 235}$$

$$\gamma_{M0} = \gamma_{M1} = 1.15 \qquad \text{for steel S 275}$$

$$\gamma_{M0} = \gamma_{M1} = 1.20 \qquad \text{for steel S 355.}$$

To make the value of the partial safety factors for structural steel in existing structures more precise, we have used in [3] the γ distribution of strength characteristics of steel for which the values of γ_M have been determined. Using the statistical characteristics of the yield stress of steels produced in Slovakia presented in [3], the following values for the partial safety factors of structural steel were obtained:

- For $5 \leq t_r < 20$ years $\gamma_M = 1.10$ for steel S 235
                                  $\gamma_M = 1.15$ for steel S 355

- For $t_r < 5$ years $\quad\quad\quad \gamma_M = 1.05 \quad$ for steel S 235
$\quad\quad\quad\quad\quad\quad\quad\quad\quad\quad\quad \gamma_M = 1.10 \quad$ for steel S 355.

For steel S 275, no statistical characteristics are available given in the very limited use of this kind of steel in our conditions. If values are needed, we recommend using the values of partial factors applicable to steel S 355.

Experience with statistical assessment of Slovakian steels does not correspond with the European practice of using a uniform value for the partial safety factor for all kinds of steel. Our experience points to the need give different values, depending on steel quality [3].

1.3.2.6 Specification of the evaluation of existing bridge structures

The process of checking the reliability of existing bridge structures is a crucial part of their overall evaluation. By evaluation of an existing bridge structure is understood a complex assessment based on processing all available information, so that it is possible to reach the optimum most economic decision on the strategy for further maintenance, repairs and reconstruction of the bridge. From this perspective, the evaluation has a key position in the overall system of bridge management. That is why it should be as unbiased as possible, considering the current technical condition of the bridge on a scientific basis as well as on the basis of present knowledge and scientific standards to do with bridge structures. In general, there are two approaches to the evaluation of bridge structures currently being applied:

- Classification approaches
- Reliability-based approaches.

Classification approaches involve assessment of a bridge by points, based on its current technical condition. It is concerned with evaluating the technical condition of the bridge structure, which is usually taken to be a complex evaluation but without checking the reliability of its current condition. In classification approaches, this is expressed by various weighted coefficients which seek to take into account the influence of failures and damage to individual elements upon their reliability and that of the entire system. The methods of evaluation vary, ranging from the purely subjective opinion of the evaluator in the form of a classification degree for the bridge structure decided by himself/herself to sophisticated methods using various mathematical approaches that should exclude any bias of the evaluator.

In this context, a very well-elaborated system of classification evaluation of the road bridges should be mentioned. These methods, developed in software form supplemented with a catalogue of failures in a database form, represent an efficient aid to immediate assessment of the technical condition of a bridge structure. With respect to a complex evaluation of a bridge structure, however, their output need not always be totally correct, particularly those for bridges with reserves in loading capacity, which are more resistant to defects and damage to bridge elements. If the evaluator does not recognise this reserve, he/she cannot evaluate correctly the impact of a possible failure on further use of the entire bridge.

Reliability-based approaches to the evaluation of bridges look at the direct influence of the technical condition on behaviour of bridge structure on its reliability. They are most frequently used with steel bridge structures, for which the methodology of considering the effect of imperfections on the response of the structure to loadings and to the resistance of its structural elements has been elaborated most. The methodology of reliability-based evaluation was gradually developed in parallel with the development of the methodology of designing steel structures.

The fundamental quantitative and qualitative parameter in reliability-based evaluation of existing bridge structure is its live-load rating factor. The live-load rating factor is generally understood as the amount of bridge capacity used by a variable short-term traffic load. Therefore it is expressed via the level of effect of an appropriate variable short-term traffic load, either road or railway traffic, which in practical calculations is simulated by ideal load models.

Loading capacity is established for separate elements of the bridge structure as a marginal condition of reliability of the relevant limit state. From the marginal condition of the ultimate limit states $S_d = R_d$, the loading capacity is expressed in general form by the equation

$$LLRF = (R_d - \sum E_{rs,Sd,i}) / E_{Q,Sd} \tag{1.19}$$

where

$LLRF$	is the live-load rating factor of the structural element
R_d	is the design value of resistance of the structural element
$E_{Q,Sd}$	is the design value of the variable short-term traffic load effect,
$E_{rs,Sd,i}$	is the effect of residual loads acting simultaneously with the traffic load (permanent load, variable long-term load, climatic loads, brake forces, lateral strokes etc.).

With respect to serviceability limit states, the live-load rating factor of a bridge element is given by the equation

$$LLRF = (\delta_{lim} - \delta_{rs}) / \delta_Q \tag{1.20}$$

where

δ_{lim}	is the limit value of a deformation quantity of a structural element with respect to its serviceability
δ_Q	is the value of the same deformation quantity induced by service effects of the variable short-term traffic load

δ_{rs} are values of this deformation quantity induced by residual loads acting simultaneously with the traffic load (these considered only when
they are not eliminated by an external intervention, such as by camber
ing of the load-bearing structure).

By setting a live-load rating factor, the reliability of bridge structure is checked at
the same time. With respect to ultimate limit states, it establishes that the design
values of load effects corresponding to the set live-load rating factor do not exceed the design resistance of the structure or its parts. With respect to serviceability
limit states, it establishes that the effects of service loads satisfy, for the set value
of the live-load rating factor, the appropriate limit values corresponding to the criteria for the relevant serviceability limit state.

Ultimately, serviceability limit states for an existing bridge are derived from
the limit states of newly designed bridges, which are prescribed by the appropriate
design standards [12].

An important procedure with bridge structures is their regular periodic checking via inspections, whose periodicity is set by the relevant standards and content
depends on the kind of inspection (see Chapter 3). These inspections represent a
source of information on the technical condition of an existing bridge, and also
make it possible to use the acquired information to reduce uncertainties in the
basic input parameters of the process of checking reliability, by which the live-
load rating factor is also defined. So it is quite possible to use the outcomes and
knowledge gained in the theoretical analysis discussed in Section 1.3.2.5 (above)
concerning adjusted levels of reliability for existing structures.

A similar approach to checking reliability and to calculating the live-load rating
factor for the existing bridges was also adopted in the Canadian and American
[13] standards for reliability-based evaluation.

Chapter 2

Causes and analysis
of steel structural failures

2.1 Introduction

Steel load-bearing structures are normally used in large and important objects, so the failure of these structures often endanger human lives and the environment, as well as causing great economic losses. The term, 'failure', includes serious faults, damage and defects that lead to a total loss of both reliability and function, either of a part or the whole structure. One of the most important factors that can reduce the number and seriousness of failures is education in the knowledge gained in this field during engineering studies at universities.

The ability to analyse the causes of structural failure cannot be obtained without an understanding of the functions of the structure. From the relatively short history of the use of steel structures it is clear that some failures have positively influenced the development of knowledge. Though unfortunate and expensive this experiment – the assessment of capacity and usability under real operating conditions – has provided a stimulus to many developments in various aspects of load-bearing steel structures (theory of buckling, structural changes and tensions in welded structures, brittle fracture, wind resonance, etc.).

The important factors in the prevention of failures include:

- A sufficient conceptual quality in the proposal, proper design of constructional details, such as nodes that accord with the structural statics of the scheme under consideration
- Proper selection of materials and joints
- Adequate professional monitoring of the design, construction, erection, and use of the structure, together with effective communication between specialist contractors and users in the design process, as well as over the whole life-time of the structure.

In the general parts of this chapter, we mainly use the documentation in [20, 21, 22]. The analysis of individual failures showed that the causes fell into a number of groups. These groups will now be discussed in turn.

2.2 Errors in the design of structures

These errors can be divided into:
- Errors in the conceptional design and arrangement of the structure (e.g. insufficient security of dimensional stability and robustness of the structure)
- Errors in the documentation for the design and the calculations (imperfections in geological documentation, false values and positions of loading effects, ignoring important influences that can have an effect on the structures, etc.)
- Imperfections in calculating and dimensioning the structural elements, selection of a calculation model that does not accord with the real behaviour and functions of the system,
- Wrong selection of constructional materials
- Errors in solution of constructional details, discrepancies between the structural solution and static scheme
- Ignoring the requirements for maintenance and anticorrosive protection.

2.2.1 Errors in the project

The causes of about one third of failures can be found in errors originating in the process of designing. The method of design and requirements for static calculations in most cases guarantee the reliability of the structure and only in exceptional cases do they cause failure. When some influences are neglected, such as off-centre stress, off-joint loading of member systems, temperature effects, this can result in a failure. There are also some imperfections in determining the correct loading values (e.g. loading of working platforms in the load-bearing structures of heating plants). Many errors originate from designing incorrect constructional details when processing manufacturing drawings as a background for the fabrication in the bridge building shop. Also insufficient communication between designers' offices and bridge building shops can lead to errors.

When solving problems with structure for complex structures, errors occur most frequently if the chosen calculation schema does not reflect the interaction among individual parts of the structure with sufficient accuracy. The failure of the steel structure of a conveyer bridge in the area of support No. 4 is such an example (Figure 2.1). It deals with a bridge for transporting coal with direct chord lattice main

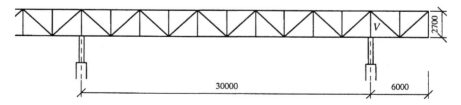

Figure 2.1 Schema for a conveyor bridge

girders having an overhanging end with spans of 30.0 – 6.0 m and theoretical height of 2.7 m.

On the basis of the inspection of the collapsed steel structure, examination of the original static calculations, drawing documentation, determination of mechanical properties of the material used in the area of destruction, it was concluded that:

- In the design of the structure, the value of the maximum axial force V vertically above support No.4 was not determined. The project did not include the design of a diagonal reinforcing frame at the place of support.
- The verticals above support No.4 were drawn in the manufacturing drawings and construction they were manufactured in the same way as other subsidiary verticals of cross-section 50.5. This part of the structure with these dimensions, was therefore underdimensioned, even for loading by its own weight.
- The failure was caused by lateral deflection of the underdimensioned verticals at the location of the bridge placement (Figure 2.2). The reason for the underdimensioning was the misunderstanding of the function and need for a diagonal reinforcing portal at the place of the girder support with the overhanging end for transfer of horizontal and vertical effects [20].

Figure 2.2 View of the collapsed part of the structure

When investigating the causes of the failure of steel structure of the supply tanks, the main load-bearing elements of the structure were assessed (Figure 2.3). Failure from the control calculation: in the design of the support structure, all members of the horizontal structure were underdimensioned while the weakest place of the system were diagonals of girders R4, where the member capacity was only

23 per cent of the design value for axial force. It was proved that the compression members of R1–R4 lattice girders were dimensioned for a simple compression without considering the corresponding buckling coefficients. Some ambiguities in the documentation the consequence that the diagonal in the outer lattice of R3 frame was not drawn in the project and it was omitted in its construction.

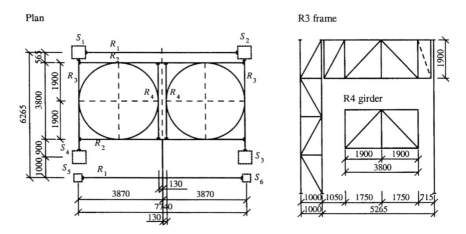

Figure 2.3 Schema for the support structure of the supply tanks

From the nature of errors, it is obvious that the design of the support structure was carried out by a team with a little experience and that serious faults in the project specification and documentation were the primary cause of the failure (Figure 2.4).

Figure 2.4 The broken support structure

2.2.2   Incorrect determination of loading

The characteristic values of the loads of load-bearing structures F_k are given by the corresponding standards, or determined by agreement among the future user, the designer and competent office.

In practice, these values, particularly those for environmental and technical loads, can reach the magnitudes considerably higher than those considered in the design of the structure.

Wind gusts have caused a number of serious failures, particularly in the masts of long-distance power lines and cranes, as well as bridge structures (e.g. the collapse of Tacoma suspension bridge in the US, which had a middle field span of 855 m). Similarly, the actual snow load (sometimes also the frost cover load) can differ considerably from the load considered in the project as regards both the size and position of the load.

The values for the characteristic permanent self-weight loads of the structure and materials specified during designing are determined on the basis of standards that take into account the dimensions and densities of materials.

Whereas in determining the self-weight of steel elements of the structure their dimensions are well-known, determining the permanent load represented by concrete and reinforced concrete elements is more uncertain, due to the fact that the manufacturing tolerances may not be met (especially in older structures) so that the self-weight of the beams often exceeds the values considered in the project by 10 to 20 per cent.

In roof structures, materials of $3.0 - 5.0$ kNm^{-3} density are used as heat-insulating layers. However, these materials are water-saturated, their density increases up to $7.0 - 10.0$ kNm^3. The actual thickness of asphalt or coarse concrete is sometimes $1.0 - 1.5$ times higher than was assumed. In refurbishment, the actual load of the individual layers has to be determined by means of test pits. To determine the actual thickness and density of a steel concrete or asphalt layer, a drill hole of at least 30 mm diameter is sufficient. When determining the weight of an insulating layer, test pits with dimensions of 100×100 mm should be prepared at ten extraction points least. On the basis of the results for the measured thickness of individual layers and density of materials (floor, roof, deck of bridge, etc.), the average value of load per 1 m^2 of every layer can be determined.

As an example of underestimation of additional loads, the failure of the roof structure of an industrial hall can be cited (Figure 2.5). The failure occurred in an area of ship roofing with a span of 24.0 m. It involved a sloping truss with parallel chords, in which a straining diagonal of 40.40.5 diameter (Figure 2.5) was broken in the third field. After the failure, the beam itself and ready area of the roof were considerably distorted.

A node cut from the point of breakage provided due to the development of the break, as well as for determining the mechanical properties of the material used.

Figure 2.5   The damaged node with a broken diagonal

From Figure 2.6 it is apparent that in the area of the roof truss examined, there was an exhaust chimney (K). In the static calculation, the possibility that sediments would gather while fabricating the forming material near the exhaust chimney was not taken into consideration nor was another additional temporary load generated by not collecting the remains of clay that accumulated during operating time. At the time of the failure, the section of the roofing in the area of the exhaust chimney was loaded by a layer of sedimented wet clay of density of $16.0 - 17.0$ kNm^{-3}.

The check calculation of the roof truss was done for the load considered in the project design and for additional temporary load generated by the sedimented clay. From the analysis, it was concluded that overloading of diagonal D_3, which originated due to the local accumulation of clay, was the cause of the failure.

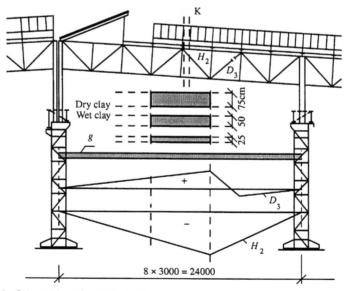

Figure 2.6   Schema for the broken roof truss

2.2.3   Inappropriate structural solutions

Inappropriate structural solutions most frequently originate from misunderstanding of load activity and the development of internal forces in individual elements, nodes and joints. The source of error can also be an insufficient knowledge of welding problems (influence of temperature on materials during welding and the origin of residual stresses in the welded joint). For example, cracks can occur in the gusset plates of welded lattice structures, but cannot occur in riveted and bolted structures. The problem of gusset plates necessitates the design of an appropriate shape of the node and selection of the appropriate quality of steel (mild steel). Figure 2.7 shows the damage to the gusset plate of a welded lattice bearer during incorrect slicing of the lower chord under strain.

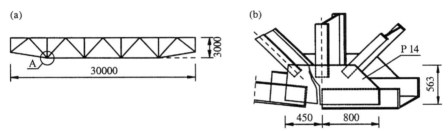

Figure 2.7   A damaged gusset plate (a) schema for the lattice bearer, (b) damaged plate in node *A*

An inappropriate proposal and incorrect evaluation of welded abutting splices of strain members can result in loss of overall capacity of the structure. The structural solution for a node in a lattice girder of the yard crane runway is drawn in Figure 2.8:

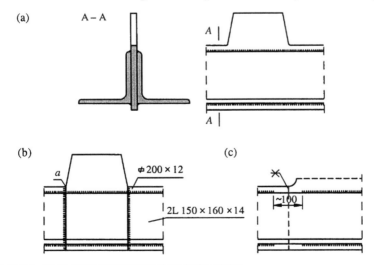

Figure 2.8   Node of the lower chord of the lattice girder

(a)   According to the project

(b)   As constructed

(c)   A correct solution.

In design proposals special attention should be paid to the transfer of the influence of bending moments on the individual members and supports at off-node load. In spite of the fact that the mode of force transfer is clear here, there are some errors in these constructional details.

Special attention should be paid both to the nodes where channels for catching water can start and to closed cross-sections. Closure of these cross-sections is not ideal, for it is often possible that water will penetrate inside, freeze and cause failures when it thaws. The types of failures caused by water freezing are sketched in Figure 2.9:

(a)   In the column base

(b)   In the closed cross-section of a lattice girder

(c)   In the bearing node of a lattice bearer

It has been found that the drain openings in these details are often non-functional, so sometimes it is better to fill in the channels with concrete.

(a)   (b)   (c)

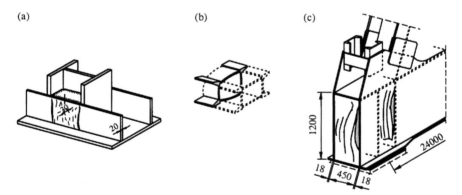

Figure 2.9 Failures caused by water freezing

Errors in structural solutions include frequently also incorrect design of stiffeners to protect the structure against a local loss of stability. In design proposals for frame corners, a simplified way of calculating the wall stiffening at the point of connection of the cross-beam with the column or the passage of the cross-beam into the column is often used.

Due to the accumulation and crossing of welds in a small area, an undesirable stress builds up in the wall caused by residual stresses, which can lead to breakage of the wall and weld material. The correct design for stiffeners of web girders is therefore very important, as in the example of girders of crane runways (Figure 2.10) where cracks can occur even after years of operation.

(a)   (b)

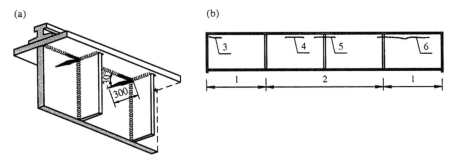

Figure 2.10 Cracks in the girders of a crane runway

Experience leads to following recommendations:

- Connection of the wall and table should be made by an abutting joint on the full thickness of the wall. For low and medium pressures exerted by bridge crane wheels, corner welds made by autogenous welding under flux are sufficient.
- Aggregation of welds should be eliminated by means of cuts in the wall stiffeners
- Wall stiffeners should be double-sided
- Longitudinal stiffeners, if they are to be part of a cross-section, should be made of steel of at least the same strength as that of the wall and they must be sufficiently spliced
- Only the unavoidably required number of stiffeners should be designed.

2.3   Defective or inadequate material

By an inadequate material we understand steels whose chemical composition, mechanical or technological properties make it impossible to make constructional elements of the required quality if they are used. The use of defective or inadequate material can result from a number of causes:

- The steel producer supplies products with hidden faults, or products that do not correspond with the order or metallurgical certificate
- The project engineer designs for the wrong type of steel
- The constructor or erection manager of the structure confuses which type of steel to use in different parts of the structure.

The yield value is an important property of steel; if steel with a lower yield value is used, its strength may be exhausted. In order to decrease the likelihood of mistakes, it is recommended that different types of steel are not used in one unit. But material with properties that do measure up to those in the project requires can nevertheless fail because of anisotropy. The variation in steel properties increases with thickness. The fork junction of a plate caused by the rolling of non-metallic drops or of two thinner plates (Figure 2.11) is a serious defect.

Figure 2.11 Fork junction of the plate caused by the rolling of two plates

Such material is not able to transfer evenly the forces acting vertically on its thickness (Figure 2.12).

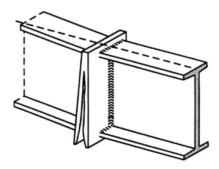

Figure 2.12 Fork junction of the front plate in the bent girder

In design proposals, it is necessary to avoid joints and details that cause stress in the direction vertical to the material surface. The steel resistance to laminar cracking is usually assessed according to the sulphur content and the amount of contraction in the direction of material thickness. In unavoidable cases it is necessary to design constructional, material and technological provisions to minimize the effects of laminar cracking. The laminar cracks that originate during welding endanger the safety of the structure. Such cracks have a step shape and develop at a depth of about one third the thickness of the plate from its welded surface.

The unfavourable effects of material anisotropy can be minimized by means of:

- Metallurgical provisions
- Check tests
- Constructional provisions.

For nodes in which there is a transfer of large loads in a direction vertical to the plate thickness (Figure 2.13), a constructional solution that secures direct transfer of the force must be designed.

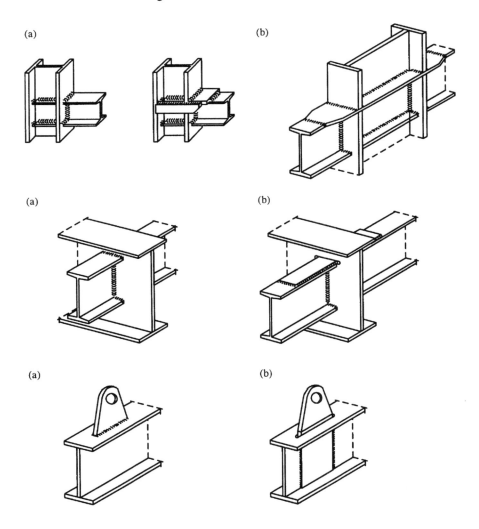

Figure 2.13 Examples of solutions for structural nodes
          (a) inadequate as regards the layering and laminar cracks,
          (b) more adequate with the use of angles to secure continuity

In welded joints it is necessary to adapt the weld or use steel which has the same mechanical properties in both directions (Figure 2.14).

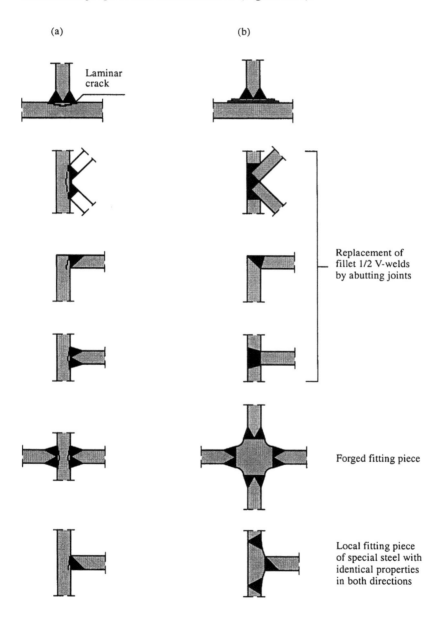

Figure 2.14   Examples of solutions of welded joints
(a) endangered by laminar cracks (b) more adequate ones

Another solution is to use an adequate welding technique (Figure 2.15) or high-strength bolts instead of welds (Figure 2.16).

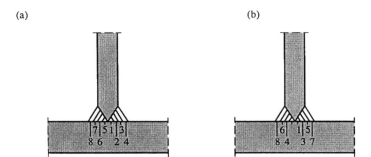

Figure 2.15   The order of welding double-sided 1/2 V-welds
(a) normally used,
(b) more adequate ones

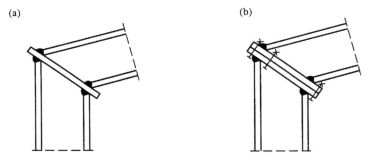

Figure 2.16   A frame corner (a) welded, (b) with a front connection using high-strength bolts

2.3.1   Steels with insufficient weldability

For welded structures it is necessary to use steels with sufficient weldability. The compatibility and required properties of welded joints are secured by prescribing correct welding techniques, choosing proper additional materials and following proper welding procedures. When the required weldability is not met, cracks can occur in the weld (caused by hydrogen embrittlement) or in the thermal zone of the weld (resulting from hardness increase). Cracks can be found in the thermal zone of welds, e.g. when welding thick plates (Figure 2.17), some cracks will form particularly in the very rigid parts of the structure, i.e. in the upper part of the column and in its base. When thick plates are connected into a rigid node, the possibility of local deformations during welding is limited.

(a)   (b)

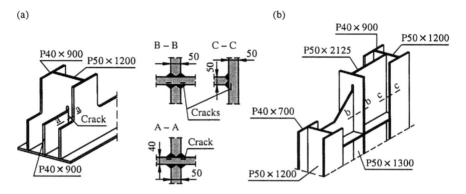

Figure 2.17 Structural details and localization of cracks: (a) base (b) upper part of the column

In Figure 2.18, the influence of rigidity of the nodes is revealed by the great number of cracks in the upper part of the column, where the thick plates were reinforced by additional elements.

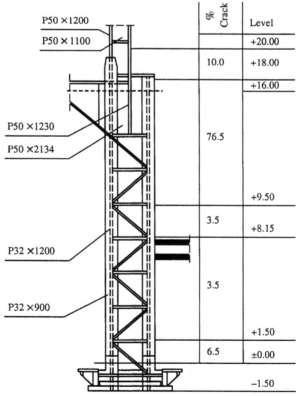

Figure 2.18 Distribution of cracks in the column

2.3.2 Steels liable to brittle fracture

Failures caused by brittle fracture have resulted in serious collapses, particularly of steel bridges and ships, which has led to extensive research of this problem. There are some known cases of brittle fracture of elements without loads from external forces; these originate in rigid structural elements where there are very high residual stresses from rolling or welding. When fracture liable steels are cut using oxygen, brittle fracture of elements loaded by high residual stresses may occur. Constructional parts can also be broken by brittle fracture at low stress (considerably lower than the design strength) when the material is not tenacious enough. Brittle cracks propagate at a high speed of 1200 m.s^{-1}, so they can cause immediate collapse of the structure. Brittle fracture depends particularly on:

- Tenacity of materials (which depends on temperature)
- Size and orientation of defects, especially planar
- Level of stress, including the influence of residual voltages
- Rate of loading
- Material thickness.

Figure 2.19 shows the failure of the left-hand part of the lower gate in the left lock chamber of the Gabèíkovo water works system [24]. This part of the lower gate consisted of a steel structure 21.95 m high and 19.28 m wide. The height of the closed cross-sections of the main girders was 2.0 m. The 20 mm thick skin plate in the upper part was of length of 3.5 m and in the lower part 3.67 m. It was made of 11.503.1 steel and in the central, most stressed area, was of length 14.78 m and made of 15.422.5 material. The rear wall consisted of three vertical 35 mm thick main girders in the lower and upper parts on the length of 4.0 m, and was made of 15.422.5 material. Similarly, 16.0 mm thick lateral walls in the central part were made of 15.422.5 steel. The main girders were joined with two horizontal binders of closed profile.

The left-hand hand part of the gate failed because of brittle failure on 20 March 1994, after 17 months operation at a loading lower than for which the two parts of the gate were designed.

When the broken parts of the gate were examined, a number of large defects and cracks were found in the vicinity of welded joints. The small 'cold cracks' originated during welding, the others from the repeated loading of the gate [25].

The cracks grew step by step, via a fatigue mechanism. After they reached a critical length, with a crack at the point of contact of the inclined wall with the table in the right main girder, a local brittle fracture originated. It was caused by redistribution of the loading of the gate and the initiation of other brittle cracks that caused brittle fracture of the whole left-hand part of the gate (Figure 2.19). The main reasons of the collapse were as follows:

- The steel structure of the left-hand part was designed appropriately and constructed in accord with the principles for statically loaded structures, but the effect of repeated load was underestimated

Figure 2.19   Brittle failure of the left-hand part of the gate in the Gabčíkovo water works

- For the most stressed parts of the gate, high-strength 15. 422.5 steel was used, which is liable to brittle fracture. This grade of steel should be welded only at high preheat ($T_p = 200 - 250°C$)
- The formation of hard structures in the steel and cracks that indicated the pressure of hydrogen gives evidence of an inappropriate welding regime (pre-heating and postheating were not carried out)
- In the welded nodes of greatest toughness, cracks developed and were en-larged by repeated stress until they reached the critical length, when they initiated the formation of brittle cracks of limited size
- By redistributing the loading to the other main girders, this process was re-peated, leading to such a decrease in the strength of the structure that the final brittle fracture of the left-hand part of the gate occurred.

2.4   Defects during fabrication and erection

Serious defects in fabrication include the wrong axial location of elements with regard to geometrical shape (off-centre connection and additional stresses caused thereby), insufficient quality of shop and erection connections (wrong fitting of parts and openings, incorrect welding technology, use of unsuitable additional mate-rials) and confusing the types of material.

Insufficient quality of welds joining load-bearing elements results from the design thickness of welds not being met, together with internal defects such as non-fused roots, pores and insufficient melting-down of edges. Faulty joints also originate from incorrect preparation of welded edges, insufficient preheating and inappropriate procedures in caterpillar laying. Damage to the material from cracks can also originate due to the grouping of a large number of welds in an area where high residual stresses arise.

Defects in fabrication can also be caused by constructional methods. Correct welding can become more difficult or impossible where there is a bad access to the site of the weld and while welding in constrained positions.

The quality of welds depends to great extent on the welder's working discipline and skills. Particularly in spatial structures made of large plates, there is a danger of distortion of plates due to residual stresses. Longitudinal shrinkage of welds and heating of part of the plate cause crimping and buckling, which is as well as structurally unsound is also aesthetically undesirable.

Erection procedures must be such that the stability and safety of the structure can be fully secured during the whole procedure. No element, connection or part should be overloaded at any erection stage. The requirements for parts stresses during erection should be checked by static calculations.

The most frequent erection defects include:

- Incorrect erection order, so that elements are erected without secured stability at given erection stages
- Incorrectly made assembly connections
- Incorrect alignment of the structure.

If erection takes a long period of time, it is necessary to take into account unfavourable weather conditions, such as snow, frost cover and wind.

Failure can occur where there is insufficient toughness of auxiliary supports (Figure 2.20).

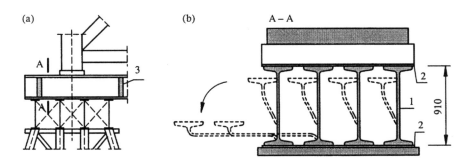

Figure 2.20 Failure of the grid under the auxiliary support (a) schema of the support, (b) transversal cut of the grid and the state after the breakdown: 1 grid bearer, 2 base made of plywood, 3 wooden enclosure

Where there are large spans (e.g. on bridges), thermal variations can cause large distortions that make alignment of erection connections difficult, especially evident in rigid spatial structures (e.g. girders of box cross-section, with trapezoidal cross-section and orthotropic flooring).

2.5   Failures caused by operations

The capacity of steel structures also depends on the length of use and on the operating conditions. The dependence of failure hazard on time is shown in Figure 2.21.

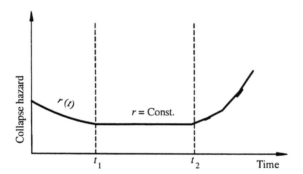

Figure 2.21 A graph showing the collapse hazard as a function of time

In the first period $(t_o - t_1)$ of use including erection, there is a greater failure hazard. The second period $(t_1 - t_2)$ of regular use, is characterized by lower failure hazard. The third period sees the structure ageing due to the long-term effects of unfavourable factors. At this stage the failure hazard increases with time. In the first two stages, the operational conditions do not have a major influence, particularly changes (deterioration) in conditions compared to those assumed in the design (change of loading, unprofessional interventions, changes in static system). The third stage $(t_n - t_2)$, i.e. ageing of steel as a constructional material involves a slow processes. The wear of steel structures over the course of long-term use is dependent on the effect of various unfavourable factors such as

- Aggressive effects of the environment and the corrosion resulting from them
- Fatigue caused by variable stresses
- High temperatures.

2.5.1   Faults in steel structures caused by atmospheric corrosion

Atmospheric corrosion – steel destruction as a result of environmental activity, affects the usable properties of steel structures. It principally affects thin-walled structures, where there is a lower reserve of material thickness than in normal structures made from rolled elements. We shall examine the effects of corrosion in order to evaluate the extent to which atmospheric corrosion leads to limit states. The majority of faults caused by corrosion of load-bearing elements initially develop without evident deformation of the structure through small losses of material from the cross-section. These losses tend to unequal, across the cross-section and result from the activity of notches.

Basic information about the effects of atmospheric corrosion on steel structures has come from research into the state of structures and the evaluation of the results of periodic assessments over a long period of time, which together give a picture of the interaction between the environment and steel structures. Non-destructive methods of evaluation are used in field surveys, and depend on the evaluator's experience, the quality standards for assessment of corrosion development and the method of estimating the level of corrosive aggression of the environment.

In the overall visual examination in the field of a structure with strong corrosion, places for probes are chosen by observing:

• The kind and quality of protective systems used
• The thickness and composition of the layers of products of corrosion
• The roughness of corroded surface
• The occurrence, depth and arrangement of corrosion pits
• Corrosion loss (residual material thickness) from non-destructive and destructive processes.

Photographic documentation of the location of parts examined (probes) and corrosive changes to structural element is done during inspection. The environment is also assessed and information compiled on the constructional and operational history of the structure. Failure of steel structures caused by atmospheric corrosion often occurs because surface finishes do not function effectively, for the following reasons:

• There were not sufficiently dimensioned or maintained
• There were applied to poorly prepared surfaces
• The constructional method used for the element or detail did not enable good application of the anticorrosive protective layers.

Increased corrosion mainly occurs in places where the constructional method or the position of an element accelerate corrosion. Examples are gaps, variations in joints and connections, cavities (in the tank) inside open profiles, restricted drying, condensation, corrosion where a column enters its base, penetration of structural elements through flooring, and soon.

In design solutions, the following faults are most frequent:

1 The use of compound cross-sections in front members with narrow gaps between them, inducing access difficult or impossible for cleaning and repainting (Figure 2.22(a) – (f)).

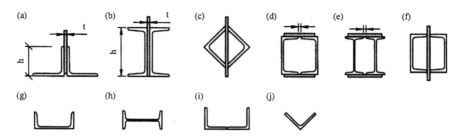

Figure 2.22   Arrangement of cross-sections

Figure 2.23 shows a better arrangement of the cross-sections of lattice members.

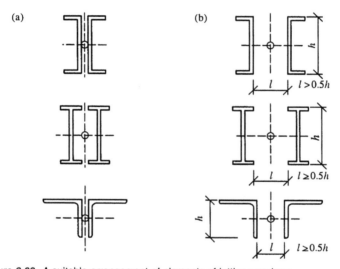

Figure 2.23   A suitable arrangement of elements of lattice members

2   Structural elements are oriented such that they provide channels to catch dirt
    and water (Figure 2.22(g) – (j)). Included here are unsuitably designed bases
    for built-in columns, without functional drain openings.
3   Use of details in places which are difficult to access for maintenance (Figure
    2.24 and 2.26); unsuitable solution for the nodes (Figure 2.25).

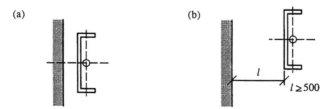

Figure 2.24   Elimination of a narrow gap

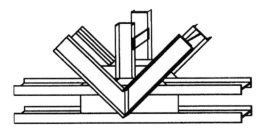

Figure 2.25 An unsuitable solution for the node of a truss bridge

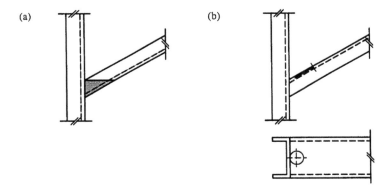

Figure 2.26 Drainage channel in the connection member

4 Designing hollow cross-sections, especially from the pipes and box, which leak when closed, are liable to humidity build-up and have cross-sections that are inaccessible for maintenance.
5 The use of intermittent welds on open structures and in an aggressive corrosion environment.
6 Building steel elements into concrete where there is high humidity, or which are filled in with material which make drying and sufficient maintenance impossible.

The amount and extent of damage caused by corrosion can be reduced by a number of methods:

- Proper selection of the location of steel structure. In cases where the locality is given, the level of corrosive aggression of the environment should be considered in the design process.
- The adoption of constructional methods that prevent corrosive damage from beginning.
- Optimizing the materials used, their combination and the anticorrosive protection systems

- Prescribing and implementing periodic assessment inspections and regular professional maintenance.

Due to corrosion, some parts of steel structures can be seriously damaged, and weakened. In cases where cross-sections are weakened by corrosion, it is necessary to asses them by static calculation. When reinforcing the parts damaged by corrosion, it is advisable not to cover such parts (Figure 2.27) because new uncontrolled deposits of corrosive products can occur there. The renovation of damaged parts can be an exception to that rule, but special attention should be paid to such parts, and anticorrosive protection should be provided carefully within reinforcing, e.g. by metallization, because in such a case it is possible to exacerbate the damage by 'bridging' (Figure 2.28). Ultimate removal of a damaged part depends on its constructional function, on the cost of removal, on the danger of further corrosion, on indication of what cracks that could occur, on aesthetic considerations.

(a)   (b)

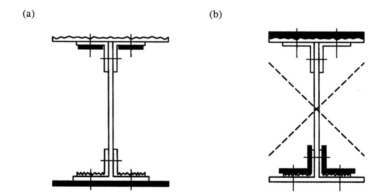

Figure 2.27 Reinforcement of a girder damaged by corrosion (a) correct, (b) incorrect

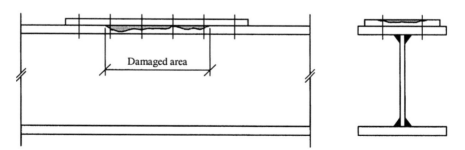

Damaged area

Figure 2.28 Reinforcement of an area weakened by corrosion

2.5.2 Damage to structures by material fatigue

The purpose of designing a structure for the limiting condition of fatigue is to ensure to an acceptable level of probability that during its design life-time the structure will not collapse or be damaged by material fatigue. When assessing material fatigue of structures, the stress in the place being assessed should consider its course over time, number of oscillations (shocks) and range of magnitude of the stresses that reflect the effect of loads on the structure during its life-time.

Material fatigue occurs in steel structures exposed to variable stress with a large number of oscillations. It occurs particularly in bridge structures, girders of crane runways and complex structures. Material fatigue can also occur in slender structures such as masts, towers, chimneys, and skyscrapers that are exposed to the dynamic effects of wind loads.

When designing and constructing steel structures subject to fatigue, the constructional details should avoid notches and abrupt changes in cross-sections. In fabrication it is necessary to eliminate defects in welded joints.

2.5.3 Effect of high temperature

The basic standards for designing steel structures are normally valid for the structures that are exposed to temperatures lower than 100°C during operation. Failure of steel structures occurs at high temperatures during fires or from the escape of combustible material. Fires that increase temperatures (above 300°C), decrease considerably the yield value and modulus of steel elasticity. The capacity of the structure thus decreases, causing large deformations.

When assessing structures damaged by fire, it is necessary to determine the extent of local and overall deformations in the various elements and parts, and to examine the effect of high temperatures on the mechanical properties of the steel. When taking specimens, start from the epicentre of the fire and then move to less damaged parts.

Permanent deformations can exceed allowable tolerances; so that actual capacity of the deformed elements must be determined by a more accurate calculation which takes into consideration the measured geometric imperfections. The structural properties of steel can be affected especially unfavourably by fast cooling while quenching the fire (decrease in notch toughness). When welding is included in renovation, it is necessary to prove the weldability of the steel that has been damaged by fire.

Parts of a steel load-bearing structure that were not irreparably damaged by fire can be renewed. In the extensive fire of a paper mill in Central Slovakia in 1995, part of a steel load-bearing structure of about 1000 tonnes was irreparably damaged. As a load-bearing structure, the mono block was divided into three expansion sections that were statically independent (Figure 2.29).

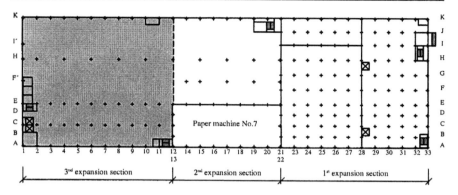

Figure 2.29 Plan of the mono block

The fire originated in the third expansion section with its epicentre in the area of transversal ties 5 – 10, and spread to parts of the first and second sections. From the results of a detailed assessment investigation of the steel load-bearing structure of the mono block, it was concluded that the steel structure of the third expansion section – diagonal ties 1 – 12 in all column orders from A – K – was irreparable damaged by the fire, so this section of the structure must be completely disassembled (Figure 2.30(a),(b)).

After the spatial stability of the structures in the first and second expansion sections had been secured, a complete assessment of all load-bearing structures was performed, and material tests and necessary adaptations were proposed.

(a)

(b)

Figure 2.30 (a) Stored paper, (b) the steel structure after fire

2.5.4 *Effects of accidental events*

Accidental events that lead to failures of structures include:
- Overloading caused by extraordinary circumstances (change of marginal conditions due to unexpected reduction of support, landslide, climatic disasters, vehicle collisions)
- Loss of strength, change in physical properties of materials (fire, large temperature changes, chemical affects)
- Seismic influences and the natural calamities
- Failures of technical equipment, explosions and other accidental events.

The failure of the masts carrying 2×400 kV electric wiring in Slovakia is a special case of the effect of an accidental event. Figure 2.31 shows the positions of masts from No.69 to 73 involved in the failure, and also the span of the conductors. Mast No.72 of RV type (II) was the fault point in the wiring track, and

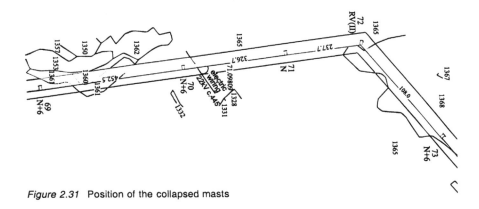

Figure 2.31 Position of the collapsed masts

as well as vertical and wind effects it also transferred the resultant of the tensile forces generated by the self-weight of the conductors.

The mast was a free-standing single-trunk quadrapod mast with varying trunk height meeting at an angle. The theoretical height of the mast was 47.2 m; the conductors were arranged in three levels above each other. The isolated mast No.72 was located at the foot of a hill near a sharp terrain break. The corners of the mast were placed on separate concrete bases embedded in the sloping land, with a height difference between individual bases from 2.0 to 3.0 m.

During the examination it was found that a considerable landslide had occurred near mast No.72. The slide had occurred at various heights on the slope above the mast. There was a significant lower landslide on the edge of slide area, and where there was a crack in the soil, there was a vertical shift of 400 – 500 mm and a horizontal shift of 150 – 200 mm (Figure 2.32). In the near vicinity of the upper corner of the mast in the direction of the electric wiring there was a 300 mm deep local landslide with crack width of 150 mm (Figure 2.33). The direction of the landslide was approximately coincident with the resultant tensile force of the conductors acting on mast No.72. The state of land surrounding the mast proved evidence that intensive farming and a large amount of rain had led to a change in the hydrological conditions of the steep land. During the assessing examination, water was pouring strongly from an underground source on the edge of the lower slide area, at a distance of 7.0 m from the lower base of the mast.

To find out the causes of the failure, a check static calculation of mast No.72 was performed. This showed that the steel structure of the mast was correctly design-

Figure 2.32 Landslide in the neighbourhood of mast No.72

ed in accordance with the standards valid at the time of construction. The tension in the corners caused by loads immediately before the collapse was 88 MPa, which represents approximately 30 per cent material utilization. From the results of the check static calculation, it was established that even at half the size of the measured slides (150 mm vertical settling and 75 mm horizontal slide) of one mast base, failure of the mast would occur. On the basis of the results of a detail assessing examination and a check static calculation, it was concluded that the cause of the failure was the landslide near mast No.72. In the failure, the steel structures of

Figure 2.33 Local landslide at the mast base

Figure 2.34(a) Collapsed mast No.72

masts at supporting points No.72, No.71, No.70 were totally destroyed and the trunk of mast No.69 was seriously damaged (Figure 2.34(a),(b); Figure2.35).

Figure 2.34(b)   Detail of mast No.72

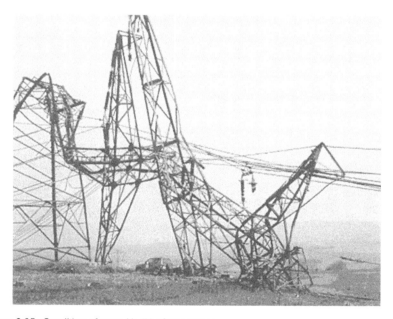

Figure 2.35 Condition of mast No.71 after collapse

In replacing the wiring, a new route was determined on the basis of a detailed geological research and new masts were located outside the slide area.

2.6   Results of defect analysis

It is obvious that the failure of steel load-bearing structures is rarely caused only by just one thing. In most cases the failure is a result of many causes acting simultaneously. In such cases the main cause is judged to be one whose influence on originating the failure is determined as the greatest. Experience shows that the greatest number of collapses are caused by erection errors, and errors in project design, as along with insufficient (or even no) maintenance of the steel structures.

In the professional literature there are many analyses of failures and their causes. As an illustration, examination of 150 analyses of failure gives the following proportions of causes:

- Defects in erection   29.5%
- Errors in the project design and shop documentation   26.7%
- Operating failures (overloading, etc.)   14.3%
- Errors during fabrication   12.4%
- Insufficient steel quality   10.5%
- Imperfections in standards and technical conditions   6.6%

With respect to technical causes of failures, analysis of the data for 63 failures gave the following results:

- Loss of stability   41.3%
- Breaking of basic material   22.2%
- Breaking of welding joints   23.8%
- Other causes   12.7%

Given the results of this analysis of the causes of failure, it is necessary to draw attention to the importance of the erection stage in building steel structures, which tends to be underestimated in the education and training of professionals nowadays.

With regard to faults in project documentation, it is worth remarking on the fact that even during engineering studies, a certain aversion of young people to dealing with conceptional solutions and attending to appropriate constructional details can be observed with a contrasting preference for static calculation using computers.

Chapter 3

Assessment of steel load-bearing structures

3.1 Introduction

Steel load-bearing structures must be operated and maintained according to appropriate standards and regulations. The overall technical condition of steel structures is monitored by regularly repeated preventive and detailed inspections. The set of methods for examining technical condition of steel load-bearing structures is comprised as an assessment. The need to assess steel structures has arisen from numerous failures of such structures, local defects in their elements and parts, and by the need to refurbish load-bearing structures after their physical life has been exhausted.

In the near future, a functional assessment should become a basic obligation for users of exposed steel structures, particularly for nuclear and thermal energy plants, in the chemical industry, for pipelines, in transportation, and any other fields where failure would have considerable negative effects on the environment, would threaten human lives and would cause major economic losses. Steel structures are sometimes assessed randomly, rather than in regular tests, and in many cases without the necessary instrumental equipment. A new procedure, designed on the basis of long-term experience in refurbishing and monitoring the technical condition of steel load-bearing structures under operation, makes assessing activities systematic, according to their purpose, unifies the methodology of assessment and specifies the necessary instrumental equipment. It presupposes the gathering and continuous keeping of documentation to do with load-bearing structures to provide the background for decision making about repairs and methods of refurbishment. Assessment should:

- Ensure (or increase) operating reliability
- Extend life
- Determine the causes and effects of particular defects
- Provide reliable background information for decision making on possible further use of the whole or a part of the original structure
- Provide background information for effective proposals for refurbishing a load-bearing structure.

3.2   Background for the assessment

During assessment, the currently applicable technical standards and regulations are followed. The older standards and regulations serve purely as background information. Before the assessment is begun, it is necessary to determine the kind of assessment and the purposes its results are to serve. It is also necessary to agree with the user on the time of the assessment (e.g. during a period when operation is interrupted).

From previous experience, it is apparent that inspections of steel structures are not performed regularly, that documentation on subsequent adaptations during operation is often not prepared and archived, and, in many cases, that their original documentation is missing. Sometimes, no basic maintenance of a steel structure has been performed either, the effect of minor damage has not been removed and protective coatings have not been renewed regularly. The documentation of the actual completion of the structure can serve as a background for all kinds of assessment activities.

The original project documentation should include:

- Working designs with static calculations
- Working drawings with static calculations of connections
- Documentation of related sections of the structure and technological equipment
- Erection and construction records of the construction contractors, as well as manuals for their technical equipment
- Drawings of the completed structure
- Protocols for acceptance of the completed structure.

The user should keep archives for documentation of any change, adaptation, damage, or repair of the steel structure.

3.3   Methods of assessment

The standard system of assessment of steel load-bearing structures is shown in Figure 3.1.

The actual technical condition of a steel structure is determined by various diagnostic methods. The classification of these methods depends on the kind and importance of the results of assessment. For identification of the nature of the structure and inspection of its geometric shape, visual non-contact and contact methods are used. Visual methods include assessing procedures in which direct contact with the monitored structure is not required. These are the basic methods for periodical assessment, which provide user with the initial information on the technical condition of the structure. They also serve in deciding the procedure, extent and method of taking measurements for the assessment.

In contact methods, appropriate measuring equipment is put in place, usually on the surface of the structure but when appropriate built directly into it. This equip-

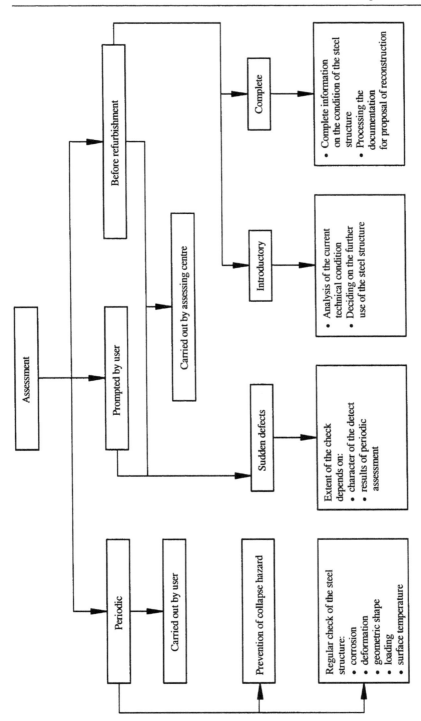

Figure 3.1 Kinds of assessing inspection

ment can monitor those physical quantities, such as shifts, deformations and opening of cracks, etc. by which it is possible to establish the technical condition of the
structure.

Destructive and non-destructive methods are used to determine the crucial properties of a material. Non-destructive methods are designed to assess both the
basic properties of material, and the extent of damage without destroying a part or
the whole of a structure. Non-destructive methods for determining material properties include hardness testing, dynamic and ultrasound methods, and radiometric
and radiographic methods. The properties of a material, in particular its strength
characteristics, are determined indirectly by means of physical-mechanical quantities such as hardness, elasticity, speed of propagation of acoustic signal in the
material tested, diminution of gamma radiation while transiting or being reflected
by the atoms of the material, etc. Destructive methods are used to determine in
particular the strength characteristics of materials by testing samples taken from
the completed structure. Using destructive methods on a structure while it is in
operation reduces the strength of the structure or part of it, so they are rarely used
on steel structures.

Among methods of assessment, a special role is played by the loading test.
This test is used on the steel structures of ground constructions, particularly steel
bridges. The method uses the results of all partial methods of assessment and together with the results of the loading test, this give a more comprehensive view of
the overall condition of the structure, particularly of its behaviour under loading.
Loading tests also provide basic information about the appropriateness and valid-
ity of the chosen model of calculation with respect to further assessment of the
structure. The tests are usually performed when new bridges are commissioned,
but they can also be done as control tests during the lifetime of the structure particularly after important interventions in load-bearing structures, such as example
when they are refurbished.

Loading tests are divided into static and dynamic kinds. Under currently valid
standards, static loading tests are prescribed for all new permanent bridges with
span greater than 18 metres. Dynamic tests are done on new bridges with span
greater than 50 metres. With existing bridges, both types of loading tests may be
performed on the decision of the inspector of the bridge structure and according
to the results and recommendations of inspection of the bridge. For some bridge
structures of great importance, on the basis of results of diagnostic inspections,
long-term monitoring of the bridge's reaction to service loading may be recommended. By comparing the reaction of a load-bearing structure at the time of com-
missioning the bridge to that obtained after a certain period of use, it is possible to
assess changes in the behaviour of a bridge structure. However, this comparison
has to be accompanied by classical methods of assessment, which will then help
determine more precisely the reason for changed behaviour of the bridge.

3.3.1   Periodic assessment

By its preventive nature, periodic assessment helps to preserve operating reliability, decrease maintenance costs and extend the life of a structure. In carrying out periodic assessment by means of preventive inspections, the assessor:

- Checks the geometric shape of the load-bearing system by measuring vertical and horizontal shifts of important nodes
- Checks that the structure as a whole does not show any distortion or excessive vibration of both load-bearing and reinforcing parts of the structure
- Determines and records local deformation of the main load-bearing parts of the structure
- Determines the condition of surface protection and size of corrosion losses
- Monitors occurrence of increased surface temperature of load-bearing structure
- Determines the condition of joints and connections of elements in a load-bearing structure
- Checks anchorage and movement of foundations, particularly settlement of pillars
- Determines deviations from the previous shape and from the static system of the steel structure
- Checks whether permitted technical and applied loadings have not been exceeded
- Checks that excessive wear of exposed parts has not occurred (e.g. crane rails or the tyres of crane wheels)
- Monitors occurrence of cracks in load-bearing structures and welds.

By detailed periodic inspection, the overall physical condition of load-bearing structures and that of associated parts is determined. The scope, depth and period of time between periodic inspections are determined according to function of the structure, operating conditions, condition of ground soil, and possibility of occurrence of failure in general.

To determine in the period of time between inspections in periodic assessment programme, new structures can be classified into the following groups:

Group A
- Spatially complex and exposed constructions and technical structures, such as nuclear power plants, chemical plants, pipelines and distribution systems (crude oil, gas)
- Steel bridge structures and conveyor bridges
- Structures for passenger cable ways
- Load-bearing structures for technical equipment in hydraulic constructions with moving parts

- Aerials and television towers, masts, load-bearing structures for radio telescopes and radar
- Other functionally similar load-bearing structures.

Group B
- Industrial, civil, agricultural and residential single- or multi-storey constructions
- Technical load-bearing structures for blast furnaces, boilers of thermal power plants, mining, drilling derricks, and cooling towers;
- Masts for electricity lines, lighting towers and masts up to a height of 25 metres
- Girders of crane runways.

Group C
- Service and technical parapets;
- Staircases and railings
- Roofing, light wells and other types of auxiliary structure.

In general, preventive inspections are performed as follows:

At least once in six months (usually in spring and autumn) for structures that are extraordinarily dynamically loaded or for which it is necessary to make sure that the direction, height or other requirements due to operational reasons have not changed, and on structures located on ground where there is sub-surface activity on undermined territory (mining, underground ways, etc.)

At least once a year (usually in spring) for other structures in Group A, according to operating conditions;

At least once in five years for structures in Groups B and C, according to their operating conditions.

If faults and defects, that could lead to more serious threats to operation or safety are detected during preventive inspections, a detailed diagnostic inspection should be performed immediately.

A detailed inspection of the structure should be performed, even when there are good results from the preventive inspection, in the following time periods:

At least once in three years for steel structures that are extraordinarily dynamically loaded, as well as structures on ground where there is sub-surface activity

At least once in five years for other structures in Group A;

At least once in ten years for steel structures in Groups B and C, according to operating conditions.

Periodic inspections should be performed by skilled agents who are able to identify defects and determine their seriousness. If the identified faults and defects threaten the reliability or operation of the structure, they should be dealt with as soon as possible.

Checking inspections of coatings and other kinds of protection against corrosion are performed at periods corresponding to the degree of pollution – the corrosive aggressiveness of the environment where the structure is located. The results of periodic assessment are continuously recorded by the user in assessment book accompanying the checking equipment. The primary record will be the initial condition on completion of construction. If the records of the condition after completion of construction have not been preserved, it would be necessary to prepare a primary record upon the first assessment.

For bridge structures there is a systematic procedure set out precisely by the corresponding standards and regulations, regime and content for preventive assessment. This is carried out within the supervisory activities for bridges. These include regular inspections, divided into:

- Current
- Checking
- Detailed or main.

Current inspections of bridge structures are performed by bridge administrators. Their periodicity is usually determined by the type of bridge. With bridges for ground communications, current inspections are done twice a year. They should always be done before and after winter (i.e. in autumn and spring). Railway bridges are subject to current inspections once a year, as they are not influenced by winter maintenance of roads. In current inspections all accessible parts of the bridge are assessed visually. Slight deformations are rectified immediately after inspection, in the course of regular maintenance of bridges. Deformations and defects to greater extent and importance are notified to supervising bodies that ensure repairs carried out in accordance with the repair and reconstruction schedule.

Checking inspections are carried out once a year and are supervised by the superior authority to the bridge administrator. Detailed or main inspections (for railway bridges the term, detailed or revision inspection, is used) in Slovakia are carried out once in three years on railway bridges and once in four years on ground communications. The detailed inspections of railway bridges are performed by auditing teams, which are responsible to diagnostic centres established in the Bridge districts of Slovak Railways. Ground communication bridges are checked by diagnostic centres of the Slovak Road Administration.

The task of detailed inspections is to assess both visible and covered parts of bridge structures with respect to their technical condition and its impact on the reliability of the bridge and its operation. Decisions about further serviceability of the bridge, or the need for temporary interventions into the bridge structure, or its reconstructional result from the detailed inspection. Due to the importance of detailed inspections, the appropriate body in the state administration is entitled to invite some other participants to the inspection. These can be experts from universities or research institutes.

3.3.1.1 Checking geometric shape

When checking the geometric shape of a steel supporting structure, the position of selected nodes in the horizontal and vertical planes are measured. The main nodes of the system are the decisive points – intersection points of axes of diagonals and cross-beams with column axes. These points can be specified by fixing permanent marks that facilitate the assessment of periodic measurements. Geometric shape is checked by locating the precise position of these selected points over prescribed periods of time, using geodetic measurement. The results are recorded on a schematic diagram of the structure, indicating the date of measurement, temperature of the environment at the time of measurement and signature of the person in charge.

As an ultimate limit for the horizontal shift of the node from both dead and incidental long-term loads, the following value can be considered:

$$v = (\frac{1}{500} \div \frac{1}{1000}) \cdot H \tag{3.1}$$

where H stands for the vertical height of the measured node of the structure from its anchorage. Ultimate values for vertical shifts originating during operation are prescribed by standards.

When judging deviations, it is necessary to use either the original or the adjusted geometric shape. When the standard values for horizontal and vertical deformations have been exceeded, it is necessary to obtain an expert opinion on the effect of increased deformation upon reliability of the structure. The horizontal shifts of nodes are determined with accuracy of ± 5 mm or ± $H/2000$.

When checking the geometric shape of spatial systems with large spans, photogrammetry can be used effectively. This method was used to determine the actual geometric shape of spatial prestressed suspension cable systems [31], as a background to checking the structure and its further rectification by means of additional prestress. A view of a cable pipeline bridge with marked measured nodes is given in Figure 3.2.

Figure 3.2 Measurement of the shape of a spatial prestressed cable pipeline bridge

The location of nodes – places of connections of both load-bearing and wind suspensions to load-bearing wind cables (Figure 3.3).

Figure 3.3   Measured nodes on wind cables L_7, L_8 and L_{18}

Inspection of the geometric shape of bridge structures is very important. Change in both height and direction locations of a load-bearing structure can have a great impact on traffic flow on a bridge, or eventually under the bridge, with greater or lesser influence on its operation. The correct height and direction locations of the load-bearing structure of railway bridges, where even slight deviations can cause restriction of railway operation, is of great importance.

The actual height and direction location of steel load-bearing structure of a bridge is checked in the detailed (main) inspections of bridges. A longitudinal section of individual main girders is measured by very accurate levelling. The measured section is entered into the graphical record and by comparing it to the previous one, the changes in height locations of load-bearing structure can be checked.

The direction location of a load-bearing structure is also determined by geodetic methods. The longitudinal axis of a load-bearing structure, to which the location of the longitudinal axes of main girders refer, and the axis of communication bridge are measured. The measured values are recorded graphically and the changes in direction location of the load-bearing structure are established by comparison with the previous record.

3.3.1.2 Local deformations

Local deformations are more frequently induced by local overloading, thermal effects or by the impact of elements of steel load-bearing structures. If the surface temperature at places in a load-bearing structure exceeded 100°C during operation where the cause of the thermal effect was not previously taken into account, it would be necessary to determine its cause and then remove it. When an effect of high temperatures on a structure is detected, it is necessary to examine whether the mechanical properties of steel have altered. During inspection, it is necessary to determine the cause and to record the nature, shape and scope of deformation, including effects of corrosion.

The occurrence of local deformations should be documented by a sketch and photograph, and should be entered into the assessment records. The effects of local deformations upon loading capacity and safety of the element and load-bearing structure as a whole should be judged by an expert on steel load-bearing structures, and a proposal for any necessary adjustments should be submitted. Great attention should be paid to local deformations of compression members and slender walls. Here, the local deformations are manifested by geometric imperfections that significantly affect the resistibility of a given element. When such defects have been found, it is necessary to measure accurately the development and position of local deformation of the element in order to determine its shape, curvature and magnitude of its deviations from the appropriate axes of the element. In compression members, it is possible to measure local deformations as deviations from the correct position that is fixed by a tension cable. In compression walls, where the local deformation is flat, it is necessary to apply geodetic methods in which the area being measured is replaced by a regular network of points marked graphically on the wall (by point, chalk, etc.). The local deformations caused by mechanical shock should also be monitored for the possible origin of cracks. During the Second World War, the upper chord of the latticed main girder of the Mária-Valéria bridge over the Danube was damaged (Figure 3.4).

Figure 3.4   Local deformation of the upper chord of a lattice bridge

During assessment, it was found that 25 per cent table widths and chord angle were damaged. During refurbishment the cross-section of the upper chord was reinforced in this place (see Chapter 5). Another similar case of local deformation was recorded in the inspection of the bridge across the river Little Danube near Kolárovo [35]. The local deformation of the compression vertical was caused by a car crash and seriously endangered the reliability of this element. This led to a decrease in the loading-bearing capacity of the bridge (Figure 3.5), because in the accident a part of the vertical cross-section was broken. The elements of over bridge longitudinal reinforcements are frequently subject to damage by car crashes (Figure 3.6). In such cases, checking the straightness of the upper chords of lattice girders is required because the deformation of reinforcement can cause deviations of these elements from the plane of main girders [36].

Figure 3.5   Local deformation of the vertical of a lattice bridge

Figure 3.6   Local deformation of members of the over bridge floor longitudinal reinforcement

3.3.1.3 Condition of surface protection and measurement of corrosion losses

When the condition of the coating of a steel load-bearing structure is being checked, it is necessary to focus especially on locations of frequent occurrence of corrosion, such as the bases of columns, where columns pass through floors (in bridges these are places where web members pass through the bridge floor or road), and locations where humidity, dust and other impurities usually accumulate.

When 30 per cent or more of the surface protection is damaged by corrosion, the coating system should be fully renewed. Corrosion loss gradually reduces cross-sections of load-bearing elements, so it is necessary to monitor its extent regularly. Corrosion losses must be monitored around the entire circumference of the cross-section.

At the measured points, the surface should be free of corrosion products down to sound metal. Rough dirt should be removed by a scraper and wire brush. The surface should then be ground finely, e.g. by an electric grinder. It is possible to measure the thickness of sound metal by a vernier gauge. A digital ultrasound thickness gauge is of general application. The results of measurements are then recorded in the records of measurement, which should include:

- Date of measurement
- Kinds of measuring instruments used
- A diagram of measured locations (probes)
- Method of cleaning the surface
- Measured values of thickness of sound metal
- Evaluation of measurement results.

There must be at least three measurements of thickness per measured location. For calculations the average thickness of the value t_{pr} should be used:

$$t_{pr} = t_a = \frac{\sum_{i=1}^{n} ti}{n} \tag{3.2}$$

where t_i is the measured thickness and n is the number of measurements.

Where corrosion reduces the thickness of a monitored element by more than 5 per cent, it is necessary to make a judgement in static terms of the effect of weakening due to corrosion in reducing the loading capacity of the element.

During detailed diagnostic inspection of the bridge in Figure 3.7, [32], being considered for refurbishment, the most dangerous place with respect to corrosive damage were the places where the web members of the main girders passed through the reinforced concrete pavement slab. The corrosion losses on the P.16 500 mm vertical plate measured at a distance of 50 mm from the edge reached as much as 50 per cent, while at a distance of 100 mm, the corrosion loss was 36 per cent.

Figure 3.7 Location of measured places on the vertical

The measurement of corrosion losses using an ultrasound digital thickness gauge is shown in Figure 3.8(a), while a frontal view of the place of weakening is shown in Figure 3.8(b).

(a)   (b)

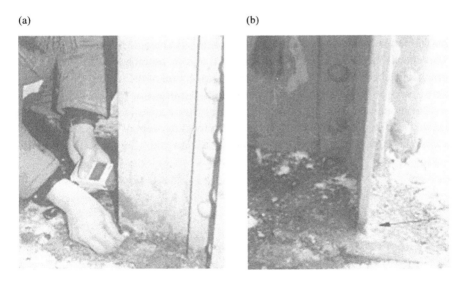

Figure 3.8 (a) Measurement of corrosion losses, (b) a front view of the weakened area

The other sensitive places for corrosion are the gaps between members joined by connecting rivets. Here, there can be a gradual opening of gaps so that local gap corrosion occurs (Figure 3.9).

Figure 3.9   Opened angles of the diagonal

3.3.1.4 Checking of joints and connections

Older structures were usually riveted, newer ones are welded, with welded or bolted erection joints. Exposed details, frame corners, connections in nodes, neck welds and rivets of girders need to be checked.

In *welded joints*, checking of welds should focus on the occurrence of cracks in welds and in the base material in the vicinity of welds. Random checks should be performed within a range of at least 10 per cent of the length of supporting welds. The checks are done visually, but a magnifier is recommended. If there are any cracks, it is necessary to perform a detailed inspection of the vicinity of the cracks using more accurate methods, e.g. penetration or ultrasonic tests.

During construction of an office building with a steel load-bearing structure (Figure 3.10(a)), a crack was found in the connecting weld of the I-section diagonal at the node of the lattice girder in the roof structure (Figure 3.10(b)). It was

(a)

(b)

Figure 3.10   (a) Roof structure of a building, (b) a through crack in the connecting weld

established that the weld was made without sufficient prior treatment of the edges of the material along the whole length of the weld.

Bolted and riveted joints   When checking bolted joints and connections, attention should focus on their completeness and on loose nuts and missing washers under nuts. Missing bolts and rivets must be replaced as soon as possible. Their quality must match to that of the original connecting devices. The length of the bolt body must be such that after tightening the body would overhang the nut by at least two threads. Loosening of bolts and rivets can be checked visually for by tapping. The nuts of loose bolts must be tightened. Bolts should be tightened only by standardized wrenches without extension levers. Loose and damaged rivets must be replaced. In the inspection of the bridge in Figure 3.11 [32], a few loose rivets were found, which were there replaced during refurbishment.

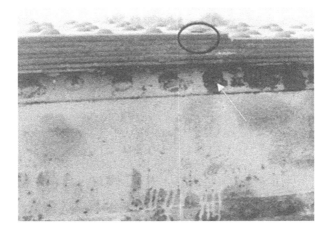

Figure 3.11   A loosened rivet in the table of the upper belt of the bridge

Checking of joints with high-strength (HS) bolts focusses on the risk of loosening of bolts, the pushing of washers into jointing material, and on mechanical damage to bolts. By random inspection at least 5 per cent of the total number of high-strength bolts should be checked. When the tightening torque of a high-strength bolt of a joint is found to be lower than prescribed, the tightening torque of all bolts in the given joint must be checked. Loose and damaged high-strength bolts must be replaced. During checking of joints, it is necessary to pay greater attention to joints that would cause increased risk of corrosion.

3.3.1.5 Deviation in the geometric shape and static system

In the initial assessment, it is necessary to check if the actual arrangement of the main nodes of the structure and the structure as a whole corresponds to the designed static system. Deviations in the geometric shape and static system are checked by comparison with the designed condition or with that found in the periodic inspections. In the inspection of a load-bearing system, it should be ascertained, whether there has been a change in geometric shape, disturbance to the integrity of the load-bearing systems and damage or alteration or removal of some elements of the steel load-bearing structure. A detail modified without authorization, and missing or seriously damaged elements can threaten the safety of the entire structure. Identified deformations should be removed by repairing the structure, returning it to its original condition, or on the basis of an expert appraisal.

The steel structure of a coal cleaning plant [33] was built in 1937. In 1978, while it was being assessed before a planned reconstruction, it was found that due to inexpert interventions the integrity of the load-bearing structure had been impaired, resulting in a change to the static system. In many places, the diagonals of lattice reinforcements were removed gradually (Figure 3.12(a), (b)) which had led to undesirable oscillations of the structure.

(a)

(b)

Figure 3.12   (a), (b) Inexpert interventions in a load-bearing system by removal of diagonal reinforcement

3.3.1.6   Inspection of origin and propagation of cracks

Cracks in the elements of steel structures are one of the most basic kinds of damage. They originate from mechanical damage caused by impacts on the element or by fatigue processes in dynamically stressed structures. Fatigue cracks are responses at places of stress concentration to operational loading. These are very dangerous kinds of damage because fatigue cracks are difficult to discover using current methods of assessment. Therefore the preventive assessment should focus on checking selected locations on the structure where fatigue cracks occur most frequently. In old steel-riveted bridges, such places are as follows:

- Riveted connections of ledgers to binders (Figure 3.13(a))
- Upper tables of ledgers under sleepers (Figure 3.13(b))
- Connections of filling members to lattice main girders (Figure 3.13(c)).

(a)   (b)   (c)

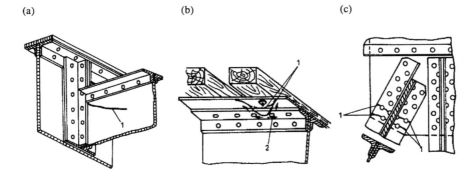

Figure 3.13 Locations of frequent occurrence of fatigue cracks in steel riveted bridges

In welded steel structures that are dynamically stressed, typical fatigue details are shown in Figure 3.14. These are as follows:
- Where there is change in the cross-sections of tension elements
- The beginning or ending of welds
- In the neighbourhood of welds oriented crosswise to the direction of main tension stresses

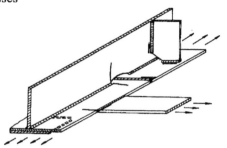

Figure 3.14 Details of typical origins of fatigue cracks in welded bridges

A place where fatigue cracks frequently occur is in the vicinity of welds connecting the longitudinal reinforcements of closed cross-sections to the floor plate in modern plate-wall bridges (Figure 3.15).

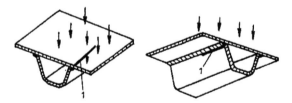

Figure 3.15 Locations of crack origins in the plate of orthotropic floors

Fatigue cracks also originate in incorrectly designed details, such as of unsuitably designed connections of transverse and longitudinal reinforcements (Figure 3.16(b)).

(a)   (b)

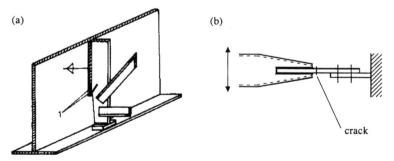

crack

Figure 3.16 Faulty constructional adaptation as the origin of fatigue cracks

Greater attention should be paid to such details in the devising the preventive assessment of steel bridge structures, such as by scheduling more frequent inspections. When looking for cracks, other than by standard visual checks, a magnifiers should be used, or ultimatelly a penetration test or defectoscopic check. Any cracks found should be recorded in the inspection report, its seriousness noted and any necessary provisions performed immediately. These should aim at completely halting propagation of the fatigue crack, most frequently by drilling off or welding in and grinding.

3.3.1.7 Checking of permitted loading

During operation and with changes in technology, increases in loading parameters or changes to loading positions can occur. The permitted applied loading can also be exceeded during erection and overhaul. Due to maintenance and repairs to roads that add more surfacing layers, the loading is often changed. Changed operating conditions are usually the source of changes in operating loading in transportation cases. Most of these cases involve change of operating speed, adaptations to railway capacity, or changed operating conditions caused by maintenance of railway tracks and stations or by cessation or restriction of operations on certain transport links.

Periodic assessment should determine:

- Whether permitted values of applied and technical loads have been exceeded
- The occurrence and location of loading effects
- Changes in technical load due to technological innovation
- Whether the labels indicating permitted loading have been damaged, removed or changed.

During assessment of the structure according to [33], excessive loads on the platforms (Figure 3.17) that were not necessary for the technical processes in the coal cleaning plant, were registered.

Figure 3.17 Excessive load on the platforms of a load-bearing structure

The results of periodic assessment are continuously recorded in the assessment book for the given steel structure. The primary record in the assessment book is an entry for the initial condition after completion of construction.

If the record of the initial condition has not been preserved, the record for the first assessment will be considered to be the primary one. In such a case it is necessary, in addition to the data determined by periodic assessment, to establish:

- Shape and dimensions of supporting elements;
- Static properties of nodes (strength and durability of connections, state of seating on foundations or on other structures);
- Basic mechanical and technical properties of materials used.

3.3.2 Assessment prompted by user

Assessment of a structure prompted by user, after discovering a serious failure of the structure or its parts, is carried out by a specialised assessing work unit.

The causes and extent of damage should be determined, with special attention focused on the vicinity of the damaged parts of the structure. The results are recorded in the assessment book. Following an accident, measures should be taken immediately to prevent further damage and secure the reliability of the remaining parts of the structure.

3.3.3 Assessment before refurbishment

When making a decision on refurbishment, as a rule a detailed inspection of the load-bearing structure should be performed. Usually, a detailed analysis of the technical condition of the supporting structure is necessary, which can verify the scope for efficiency of using original structure or its parts for the new purposes. After reaching a judgement of the physical condition of the steel structure, the results of periodic assessment will be used. The scope of the assessment prior to the proposed refurbishment should be adjusted according to the technical condition of the steel load-bearing structure and the user's intention. The assessment before refurbishment is divided into:

- The initial assessment
- The full assessment.

The initial assessment is performed while user is making preparations for refurbishment, or for changing the technical profile. It should provide answers to questions about, whether it is possible to make use of the original steel load-bearing structure and to what extent new structural arrangements are to be needed. The following documents can serve as a source of information on the condition of load-bearing structure:

- Original or modified project documentation
- Statements and reports from maintenance and operating workers

- Operation journals
- Revision books
- Records of periodic assessment, or the assessment book
- Records of repairs, modifications and refurbishment
- Inspections of the structure.

Before the initial assessment, it is necessary to establish any new requirements for the supporting structure that will result from refurbishment or technical changes.

Full assessment follows the initial assessment but before and during preparation of the design for refurbishment. The initial assessment will frequently have been carried out during operation, when some parts of the structure were not accessible (such as the load-bearing structures of thermal power plants, load-bearing elements of bridge floors, etc.). The aim of full assessment is to obtain complete information on the steel load-bearing structure so the results provide a sufficient background for effective preparation of the design for refurbishment. The differences between an initial opening and a full assessment before intended refurbishment is shown in Figure 3.18. The initial assessment was done during operation (the first part of Figure 3.18) while the full assessment was done after dismantling the boiler (the second part of Figure 3.18).

Figure 3.18 The result of an initial and a full assessment (a heavy corrosion of the boiler)

3.4   Instruments and devices for assessment

Instruments and devices used in assessing steel load-bearing structures are classified according to their purpose:

(a) Instruments for measuring the geometric shape of the structure
(b) Instruments for measuring wear and damage to the structure
   • material and joint defects
   • corrosion losses
   • surface protection of the structure
   • changes in the structural condition of the material
(c) Instruments for determining the static and dynamic behaviour of the structure
   • measurement of temperature conditions of the structure
   • measurement of stresses and deformation
   • measurement of vibration frequency and amplitudes, and their behaviour over
     time
   • working out the experimental loading.

The methods and purposes of various measurements and most important in-
strument equipment needed are described in Table 3.1. While particular types of
instruments are given in the table, it is necessary to bear in mind the rapid development of current measuring technology.

3.4.1   Measurement of the geometric shape of a structure

The purpose of measuring geometric shape is to obtain the background data needed for verifying the behaviour of the structure, taking into consideration the influ-

Table 3.1   Instruments and equipment for assessment

Measurement aim	Type of assessment	Purpose	Methods	Instruments
Geometric shape	Periodic assessment, asssessment prompted by user	Reference systems	Geodetic: • Intersection forward • Measuring of length (mechanical, optical, electric) • Measuring space grid of direction in rows and groups • Determining inaccessible distances	
			Aids for mechanical implementation of straight lines	Tensioned invar wire, plumbline
			Photogrametric methods	Phototheodolite
				Accurate comparator
				Recording device

	Preventive assessment, assessment before refurbishment	Position of constructional parts	Measuring distances, space grids for determination of inaccessible distances, intersection forward	Steel gauge with millimetre division, theodolite electronic distance finder
	Assessment prompted by user, Assessment before refurbishment	Shape and dimensions of constructional parts	Dimension measurement	Invar bands, vernier micrometric bolts, ultrasound thickness gauge
			Measurement of gradients	Shim gauge, level with micrometric bolt
Wear and damage of structure	Periodic assessment, Assessment prompted by user, assessment before refurbishment	Occurrence and dimensions of cracks	X-ray radiography	Portable X-ray machine 150 – 300 kV
			Radio-isotopic radiography	Radiation head IR 192
			Ultrasound deflectoscopy, magnetic detection of cracks	Ultrasound deflectoscopy, hand electromagnet

Measurement aim	Type of assessment	Purpose	Methods	Instruments
Wear and damage of structure			Inductive detection of cracks	Sources of current for activation of induced field
			Capillary methods	
	Periodic assessment assessment prompted by user	Corrosion loss of thickness	Ultrasound measurement of thickness	Ultrasound thickness gauge
	Periodic assessment	Thickness loss of anticorrosive layer	Magneto-strictive measurement of thickness	Thickness gauge for non-metallic coatings
			Measurement of thickness by whirling current	Thickness gauge for metallic coatings
	Assessment prompted by user	Detection of defects in structure of steel	Destructive	Laboratory tests: • mechanical • chemical • metallographic
			Non-destructive	Portable hardness tester
Static and dynamic activity of load-bearing structure	Assessment prompted by user	Deformations	Mechanical	Watch deviation meter
		Unit distortion (stress)	Electrical • rheostatic • inductive	Tensometric apparatus
		Forces	Mechanical Electrical	Dynamometer
		Amplitude and frequency of oscillation	Electrical	Dynamic tensometric apparatus, 6 or 12 channel with oscillograph

ence of alterations to its geometric shape. Measurement results can be used in proposing kinds of refurbishment, including fabrication of new additional elements and parts.

The instruments for measuring geometric shape are used to determine:
- Location in the space of structural parts
- Shape and dimensions of individual load-bearing elements and parts.

The position of individual parts of a structure can be established by direct measurement of distances along directional axes or by measurement of distances of inaccessible parts of the structure from points fixed on a directional axis. Distances of inaccessible parts can be measured using special optical and electric instruments (see Table 3.1).

3.4.2 Measurement of wear and damage to structures

The effects on steel structures of operating and of the environment are revealed over the course of time, notably by the following:

(a)   The development of mechanical processes in material exposed to the effects of long-term or repeated stress, as well as adverse temperature conditions (high temperatures extremely low temperatures). These can be manifested over time as cracks that reduce the safety and serviceability of the structure and its operating life. Similarly, plastic deformation caused by stretching of material, particularly at higher temperatures, or fatigue cracks caused by vibration of the structure, are other mechanisms of structural disturbance.

(b)   Reduction of the thickness of a constructional material by corrosive processes so that, cross-sectional dimensions are reduced, thus increasing stress on them and accelerating the development of flaws.

(c)   Reduction of the thickness of protective anticorrosive layers, increasing the risk of corrosion with the consequences described under (b) above.

(d)   Due to the effects of high temperatures, though acting only temporarily, structural changes in the steel can occur (e.g. spheroidization of lamellar perlite) with consequential reduction in the material strength and thus also the ultimate loading capacity of the structure.

The diagnostic work unit should be equipped with technical instruments (see Table 3.1) suitable for:
- Looking for defects (cracks) in structural materials (magnetic and ultrasound defectoscopy)
- Measuring structural material thickness
- Measuring the thickness of protective anticorrosive layers
- Measuring (informative) of strength properties by non-destructive methods (through material hardness) and sampling for destructive tests.
- Destructive tests on mechanical properties of steels (yield stress, ultimate

tensile strength, elongation and notch ductility) need to be necessarily performed in a testing laboratory.

3.4.3   Verification of the static and dynamic behaviour of a structure

Determining the actual static and dynamic behaviour of structures can be considered after detecting a failure in a supporting structure during periodic assessment or during assessment before proposed refurbishment.

Verification of static and dynamic behaviour is carried out:

- When it is necessary to state the scope for conditions of continued use of structures that have served for many years, should defects appear whose influence on the ultimate load-bearing capacity of structure cannot be reliably determined by calculation
- When it is necessary to extend operational loading and determine the existence of additional reserves of loading capacity
- Upon detecting various errors in newly constructed structures, when the effect of such errors upon the reliability of the structure cannot be shown unambiguously;
- After completion of a large technical project but before bringing it into operation when the quality of the structure needs confirmation (by loading tests)
- When new structures are used for the first time, whose reliability has not been confirmed in practice
- To confirm the ultimate load-bearing capacity of structures after they have been reinforced, refurbished or renewed.

The purpose of the appropriate measurements is to ascertain the effects on the structure (forces, temperature) as well as the structure's responses (stress, deformation, pattern over time).

To measure magnitudes monitored by static and dynamic tests, several single-purpose devices were developed. However, current trends favour versatile devices able to measure a number of magnitudes, such as strain, force, length changes, acceleration, temperature, etc. using just one apparatus, able to record measured magnitudes by means of a recording device, and to process them numerically, using a computer.

3.4.4   Supplementary assessing equipment

Besides the three groups of instrumental equipment mentioned above, some supplementary equipment is also needed, notably equipment for:

- Transporting and moving the instruments, and returning them to a measuring centre

- Photographic documentation of measured details and structures, as well as processing them in a laboratory
- Graphic recording of measured plots (direct recording devices with 12 channels)
- Archiving of digital records
- Processing of partial records directly on site via a keyboard
- Processing of measurement data (e. g. tensometric data via personal computers or not a direct constituent part of multi-channel measuring devices.

Chapter 4

Basic methods of reinforcing steel structures

4.1 Introduction

The types and basic methods of reinforcing steel load-bearing structures are shown in Figure 4.1. From a constructional point of view these methods can be divided into these classes:

(a)   Reinforcing by enlargement of the original cross-section of elements, reinforcement of joints and connections

(b)   Reinforcing by changing the static scheme, insertion of additional members and stiffeners in order to reduce stresses in the original structure

(c)   Combined method of reinforcement, in which the static scheme is changed and cross-sections of individual elements are enlarged.

In some cases, reserve load-bearing capacity can be found in the original structure, or loading can be reduced in order not to exceed the limit state.

In checking static calculations, two cases should be distinguished:

- Where elements (of a structure) are disassembled during reinforcing, or relieved of permanent or temporary loading; here the calculation does not differ from that for a new structure
- Where there is reinforcing of elements (of a structure) that are partly loaded, where the parts that are being reinforced are already in a stressed state, and where the mechanical properties of the reinforcing material is different from those of the reinforced material.

The methods of unloading used during the reinforcing period depend on actual conditions and possibilities. Using temporary supports, struts and draw bars is a simple method of unloading. The suitability of any unloading method must be documented by calculation (when the conditions relating to the support system have been changed, the division of internal forces in the elements of system is different), by which the reliability of all elements, joints and connections of the structure is checked at any given stage.

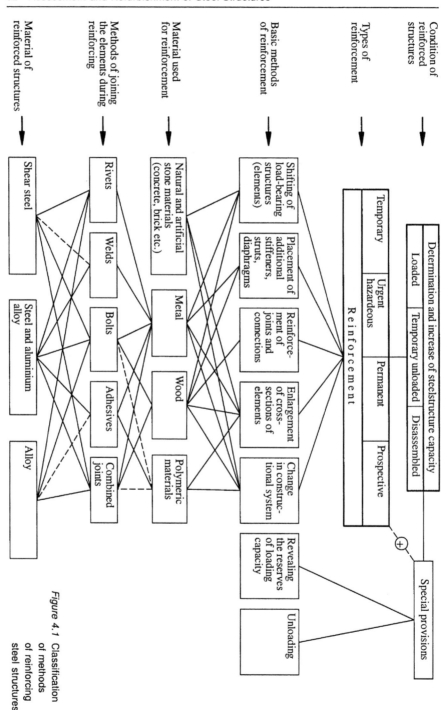

Figure 4.1   Classification of methods of reinforcing steel structures

4.2   Reinforcement of elements by enlargement of the original cross-section

Reinforcing by enlargement of the original cross-section of bent girders, members of lattice structures, struts and columns is appropriate for cases where it is necessary to reinforce only one or a few elements of the structure. The most frequently used method of cross-section enlargement is to weld new reinforcing elements to the original ones that are being reinforced. The disadvantage is that after reinforcement some residual stresses from welding remain in the cross-sections and when the material of the reinforced element is not strong enough, its liability to brittle deformation can increase. When reinforcing elements are bolted (by spot or precision bolts), it is possible that the connecting bolts do not secure a complete, tight connection of the elements and that leakly openings will allow the elements to move. This has a negative effect on the division of stresses between the original and the reinforcing elements. When frictional joints with high-strength bolts are used, the recommended value of their friction coefficient μ depends on the method of finishing the area of friction. In such cases, a lower number of bolts is required, as well as smaller dimensions for the joint or connection of the element.

4.2.1   Reinforcement of angled girder cross-sections

The cross-section of angled girder cross-sections is more economical the more the values of the moments of load-bearing capacity approach the maximum moments of flexion. When locating the new material in the cross-section, the most efficient procedure is to place the reinforcing (additive) material furthest from the neutral axis into the tabular area, thereby increasing bending stiffness of the girder (Figure 4.2(a)).

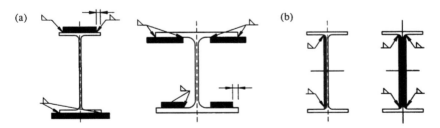

Figure 4.2   Reinforcement of angled girders

Where the effects of transverse forces are crucial, the girder wall is reinforced, leading to a reduction of shear stresses (Figure 4.2(b)). There is often no need reinforce the girder along its whole length at constant cross-section. It is sufficient to reinforce it in the area of greatest internal forces. When reinforcing angled girders, the girder height is sometimes increased, which influences the stability of the compression chord (buckling). When choosing a method of reinforcement, it is important to establish whether it is possible to unload the structure. Reinforcement is most effective when the structure is disassembled, so all reinforcing elements can be fully used and all necessary works can be performed more easily (e.g. difficult overhead welding is not necessary, but unfortunately this is often inevitable when the upper chords of built-in girders are being reinforced).

Schemes for the reinforcement of rolled I and U girders are shown in Figure 4.3. The reinforcing elements can be welded (Figure 4.3A) or bolted (Figure 4.3B) to the original cross-sections, though welding is preferable.

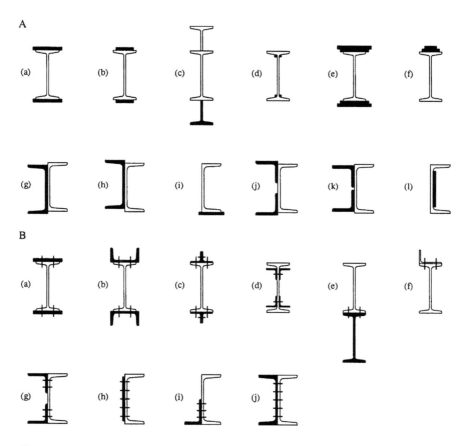

Figure 4.3 Reinforcement of rolled angled girders

The reinforcement according to Figure 4.3A(a) is advantageous because the position of the cross-section centre remains unchanged. However, girders that cannot be disassembled should be welded overhead on the upper table. This disadvantage can be removed by reinforcing according to Figure 4.3A(b) which enables good

access to the welds in the upper table. The disadvantage of the schemes in Figures
4.3A(c) and 4.3B(b), (c) is that it is impossible to extend the reinforcing element
as far as the support. Schemes for the reinforcement of combined welded girders
are shown in Figure 4.4.

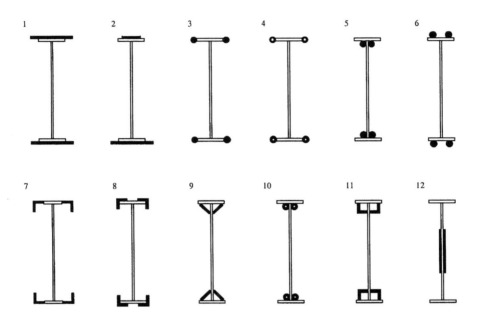

Figure 4.4   Reinforcing of welded angled girders

Welded cross-sections of girders (and the girders of crane runways) are reinforced
by welding (bolting) horizontal or inclined tables, circular and rolled profiles. When
reinforcing according to Figures 4.4 – 7, 8, 9, 11, the horizontal strength of cross-
sections is also increased. Reinforcement according to Figures 4.4 – 9, 11 also
solves the problem of buckling because it results in a closed cross-section of chords.
However, there is the disadvantage that the reinforcing elements cover the original
neck welds.

   When there is an insufficient local stability of a web-welded girder, additional
wall stiffeners are used. The use of vertical (in tall girders also horizontal) stiffeners increases the stability of the girder wall. Local swelling of the wall reduces
stability and decreases the capacity of girder wall. Where the swelling has reached
the wall area within a limit of 1/3 h (Figure 4.5), the wall can be stabilized temporarily by welding additional vertical stiffeners. In the girders of crane runways,
where the upper pressed area carries the wheel pressures, local swelling of the
wall is limited to the value of 1/10 h.

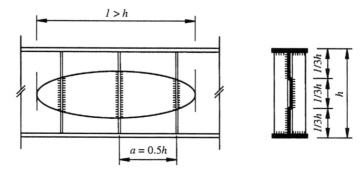

Figure 4.5   Addition of vertical stiffeners to the swollen girder wall

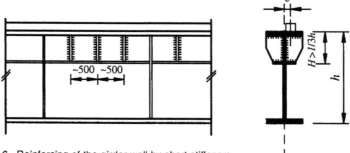

Figure 4.6   Reinforcing of the girder wall by short stiffeners

When a crane rail is centred towards the axis of a girder wall, where *e* is less than 30 mm, it is usually not necessary to modify (reinforce) the wall. At an eccentricity up to 50 mm, short wall stiffeners are welded as in Figure 4.6. Reinforcement needed because of defects in the compression zone of the web girder of a crane runway (cracks in the wall, neck welds with cracks), can be provided by means of double-sided longitudinal diagonal stiffening plates (Figure 4.7). This method is also suitable for reinforcing of the neck welds themselves.

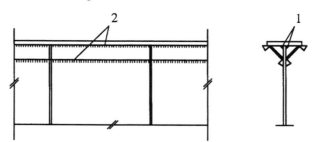

Figure 4.7   Reinforcement in the pressure zone of a web girder of the crane runway using diagonal stiffening plates: 1 location of failure; 2 welds connecting the diagonal of the stiffening plates

4.2.2   Reinforcement of member cross-sections in lattice structures

In such cases involve reinforcement of members stressed by axial forces. Where there is axial stress, it is necessary to introduce reinforcing elements along the whole member as far as the point of connection. The reinforcing elements should not change the centroidal axis of the original member, to prevent additional stresses originating because of eccentricity. In axially compressed, jointly embedded members as opposed to stressed ones, it is not necessary to reinforce the element along its whole length. The requirement to preserve the original centroidal axis remains valid as in the previous case. In compression elements, the capacity of members significantly influences buckling. Enlargement of cross-sectional area also increases the capacity of an element by increasing the moment of inertia and thereby decreasing the width – thickness ratio; because of this it is necessary to locate reinforcing elements in the original cross-section.

The members of lattice structures are reinforced by enlargement of the basic cross-section. Belt and circle profiles are welded, pipes or other rolled profiles are bolted. In Figure 4.8 some schemas for reinforcing chord or web members of lattice girders are shown. Individual cases are decided on the basis of the actual possibilities for a practical constructional solution. The profiles used for reinforcement are joined at the nodes of lattice girder and are connected to the reinforced member. The best kind of reinforcement are shown in Figure 4.8, schemas 1, 3, 9, 11 and 12. Schema 8 does not require changing the position of the neutral axis but the connection of reinforcing elements to the node is difficult; in schemas 6 and 7. In schema 14, there is a reinforcement by continuous vertical enclosure; similarly in this method, at places where there are gusset joint plates the cross-section is not reinforced, so this method can be used only in compression members when there is insufficient stability of the member in the vertical plane. The reinforcement in schemas 15 and 16 is appropriate for local defects in members. Schemas 17 to 30 show possibilities for reinforcing matched U and I cross-sections. If there are no problems with connecting, the reinforcing elements needed in schemas 17, 20 and 23 can be used.

Reinforcement for tubular cross-sections is shown in schemas 31 and 32 in Figure 4.8. Schemas 33 and 38 show a typical reinforcement of web members, but, many of the previous reinforcement schemas can also be used for web members. In a reinforcement project, it is necessary to specify the welding technique; for instance with pressed elements, the elements used for reinforcing are first welded at the nodes (if necessary) and then are welded to the profiles of a lattice girder member. In case where the centroidal axis is changed, the effect of the additional moment in the member must be assessed.

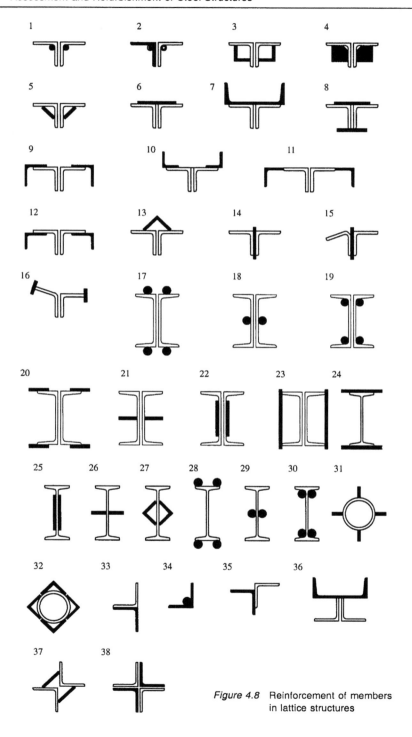

Figure 4.8 Reinforcement of members in lattice structures

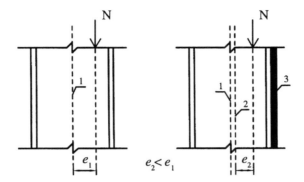

Figure 4.9   Reinforcement of an eccentric compression member:
1 centroidal axis before reinforcing, 2 axis after reinforcing, 3 reinforcing element

An element stressed by an eccentrically operating force must be reinforced in order to reduce the eccentricity as well as surface enlargement. This can be done by adding the surface to the side where there is eccentricity, so that the centroidal axis is shifted towards the axis of active forces (Figure 4.9).

4.2.3   Reinforcement of column cross-sections and struts

Reinforcement of struts and columns is done by enlarging the basic cross-section by means of chords, or ultimately rolled profiles welded or connected to the column or strut that is being reinforced by high-strength screws. Figure 4.10 shows some methods of reinforcing the struts and columns of compact, compound or articulated cross-sections. The reinforcing elements must be placed along the cross-section so that the moment of inertia can increase, which decreases the slenderness of the column or strut. The other method of reinforcement involves reducing the buckling length by means of additional stiffeners.

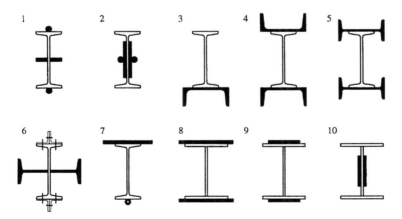

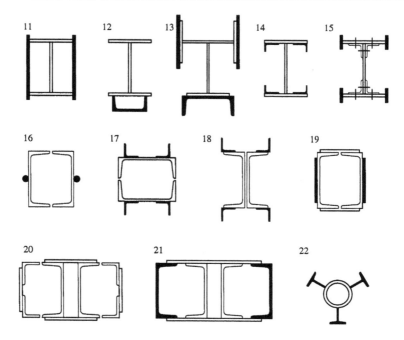

Figure 4.10 Reinforcement of the cross-sections of columns and struts

The columns with closed cross-sections (with cracks in walls or slight deformations) can be either filled or covered by concrete (Figure 4.11), and so that their capacity is increased. A considerable increase in the weight of the element is a disadvantage of this method. The capacity of built-in columns is often limited by the capacity of their anchorage, which is difficult to reinforce. Anchor bolts can be added or replaced by new tipped anchor bolts.

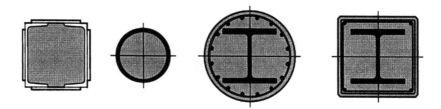

Figure 4.11 Reinforcement by concrete filling or covering

4.3   Reinforcement of joints and connections

In many cases, the joints of steel structures must be reinforced because of the structures themselves or their elements. However, in some places they must be reinforced separately. When choosing the method of reinforcement, an important role is played by the way the original joint was made (riveted, bolted, welded) as well as by whether or not the joint can be unloaded during reinforcement.

Where a riveted joint cannot be reinforced by other rivets because of constructional reasons, it is necessary to replace the existing rivets by stronger ones. Such replacement involves cutting the old rivets and drilling some more openings; additional rivets (Figure 4.12) are located according to the width of the flange at the flange angle. To divide the forces uniformly among individual rivets, rivets of the same cross-section as the original ones should be used. In Figure 4.12(b), a way of increasing the vertical flanges of a flanged angle for locating other rivets is shown. The welds connecting the scabs carry the same force as additional rivets.

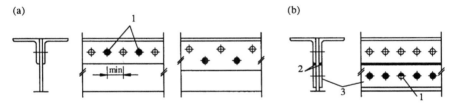

Figure 4.12   (a) a web-type girder; (b) enlargement of the vertical flanges of flanged angles squares for fitting additional rivets; 1 reinforcing rivets; 2 welded stiffening plates; 3 stiffening plates

If welded joints are faulty, for instance the welds have cracks, they should be cut off, ground and replaced with new ones. If the joints of tables or walls must be reinforced, this can be done by means of stiffening plates of the required shape (Figure 4.13).

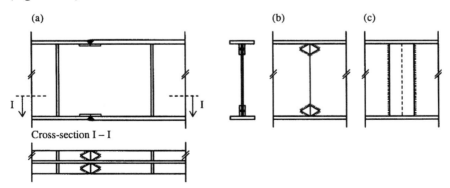

Figure 4.13   Reinforcement of connections in the nodes of tables (a) or a wall (b), (c) using stiffening plates

Reinforcement of connections at the nodes of lattice structures is shown in Figure 4.14, while in Figure 4.14(a) there is a set of additional rivets in the auxiliary connection angle place. In welded joints it is possible to attain increased capacity by enlarging the weld, gusset plate or by adding a front weld; Figure 4.14(b) shows the location of welds on the additional gusset plate.

Another possibility is to enlarge the weld by direct welding, by adding more layers (welded beads) on the loaded member. There is no consensus of opinions on this problem; in the zone of high temperatures originating in the process of welding, the basic material and welds change into a plastic state and lose their strength. In direct welding the zone influenced by welding at temperatures higher than 550°C takes on a plastic state; stresses must be re-directed into the remaining, thermally uninfluenced parts of the weld. Therefore the following conditions must be met when using this method:

- Stresses in elements in which the connection is being reinforced should not exceed 150 MPa
- On lattice girders the nodes of the lower and upper chords must be reinforced
- In order to heat as small a zone as possible, an electrode with maximum diameter of 4 mm and a 2 mm welded layer must be used
- If there is a defect in the weld (crack, cavity etc.), reinforcing must begin at that place; if there are no defects in the weld, reinforcing can begin at anyplace
- Work on reinforcing weld seams should be carried out only by highly qualified welders.

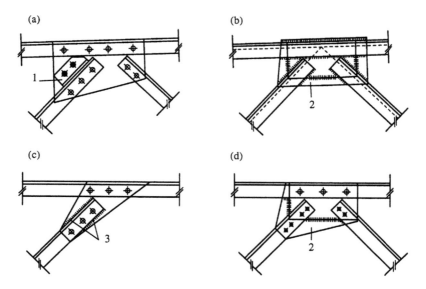

Figure 4.14 Reinforcement of riveted and welded connections in the nodes:
1 auxiliary connection angle; 2 additional gusset plates; 3 reinforcing welds

In Figure 4.14(c) there is a connected member welded all around, in Figure 4.14(d) there are reinforcing rivets in the additional gusset plate. When reinforcing riveted connections balanced co-operation between rigid welded joints and relatively flexible riveted joints cannot be expected. Therefore the welds must be designed to accommodate tensional forces that are transferred by the connection. Welding can also be used advantageously when it is necessary to enlarge gusset plates (Figure 4.14(b), (d)). When welding is used, reinforcement is considerably simplified because we can enlarge the cross-sections of smooth welded profiles easily by connecting the reinforcing elements where necessary. However, careful attention should be paid to welding techniques in particular to the internal stresses and deformations originating during welding. For the reinforcing a riveted connection by welds it is necessary to ensure good accessibility to all welds.

When reinforcing joints and connections, welding on riveted structures is to be preferred. However, this assumes good weldability of the material. This should be confirmed by appropriate tests. In many cases, especially when the joint is reinforced by ordinary bolts or rivets, it is efficient to use high bolts.

The combination of various methods of connecting (welds, rivets, precision and thick bolts) in one joint is not recommended. The strength of such a joint will be determined by whether (a) the welds or (b) the rivets and precision bolts (in precision joints) has the greater strength.

Balanced co-operation between prestressed bolts in a frictional joint with other bolts or rivets in one joint should not be assumed. Such co-operation can be expected only if the joint is statically loaded. In dynamically loaded joints, co-operation of prestressed bolts in friction joints only can be counted on with abutting joints. In such a case, the capacity of the combined frictional and welded joint $F_{b,w,Rd,}$ is given by the relation:

$$F_{b,w,Rd} \cong k_b . F_{S,Rd} + F_{w,Rd} \tag{4.1}$$

where $F_{S,Rd}$ is the designed capacity of the frictional joint
$F_{w,Rd}$ is the designed capacity of the welded joint
k_b is a coefficient depending on operating conditions; for statically loaded joints $k_b = 1.0$ and for dynamically loaded joints $k_b = 0.8$.

4.4   Reinforcement by changes to the static system

Reinforcement by changes to the static system enables a more favourable division of internal forces (as well as stresses) in the individual elements of a load-bearing structure. It is effective method of reinforcement that offers a wide range of possibilities. At the same time it is a difficult method with regard to monitoring changes to internal stresses and their effect on the stresses and stability of structural elements.

Changes to the static scheme of various types of structures can be divided into those:
- Without significant change to the static scheme; e.g. by increasing the column strength in transverse section to obtain a division of internal forces at the cross bonding of the material
- With partial change to the static scheme, e.g. by adding a draw bar, a girder is changed into a statically indefinite strut frame.

When the static scheme is changed, loading effects at points of bedding or support are also changed. Such adaptations include changing single girders into continuous girders, introducing grid systems, connecting independent elements and parts into the spatial operation of the load-bearing system, adaptation to obtain the action of a frame (e.g. changing a three-hinged frame into two-hinged one). The bedding of a structure can be changed by altering fixed kinds of bedding or changing sliding bedding to fixed kinds. Sometimes, on the other hand, the change is aimed at decreasing redundancy. Such cases include those where the supports of continuous girders have collapsed or foundations that were thought to be built in have slipped. The correct arrangement of stiffeners for securing the spatial strength is very important, particularly for plane structures. It needs to be remembered that the importance of stiffeners in older structures was often underestimated; nowadays the need for corresponding spatial reinforcement is recognized and in most cases is respected in designing new projects, as well as in completed structures.

4.4.1 Insertion of additional stiffeners and members into the original structure

Reinforcement can be done by inserting additional stiffeners that increase the spatial strength of a structure and reduce the buckling lengths of the compressed mem-

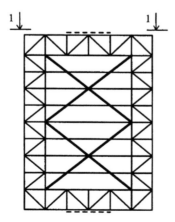
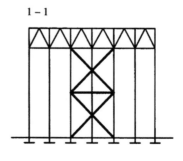

Figure 4.15 Additional stiffeners in the roof plane and in the front walls of the structure

bers. In short halls (60 – 90 m long) it is useful to provide an additional stiffener (Figure 4.15), whose strength will carry all horizontal reactions of the columns in the lower chord of the lattice girders and from the structure of front lattice stiffeners. The additional members are marked by the thick lines.

When it is necessary to increase the capacity of the lower chord members of a lattice girder, it appropriate to use additional members that shorten the buckling lengths of both chord and corresponding web members in the plane of the lattice girder. Those additional members will also reduce the effects of local bending moments where there is an off-node loading of the upper chord. In other cases when members are added, the lower chord is unloaded, or the loading is carried directly into the node of girder. Figure 4.16. shows various ways of bedding additional members in a girder node. This method is used for reinforcement of lattice trusses, tie beams, crane runways, conveyer bridges, towers and masts, particularly when reinforcement by enlargement of the basic cross-sections is difficult. Rather than reinforce weak elements, it is better to reduce their stress, in any case the insertion of new members is comparatively easy to do.

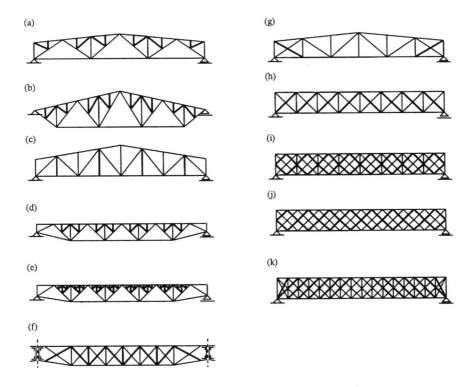

Figure 4.16 Insertion of additional members in lattice structures

When there are alterations to a roof deck, the weight of the new deck with proper isolation can be greater than that of the original one. In such cases the axial forces in the members of original girder are higher. As shown in Figure 4.16(a), (d), (c) nodes for inserting new trusses are created, and at the same time the buckling length of the upper chord plane of the lattice girder is shortened. In Figure 4.16(f), (g) – (k) the buckling lengths of the compression diagonals are shortened and the axial forces in the original diagonals are reduced in the modified lattices. In the case of refurbishment of halls, their purpose is sometimes changed when there is a need to create a suspended ceiling at the level of the lower chord (Figure 4.17).

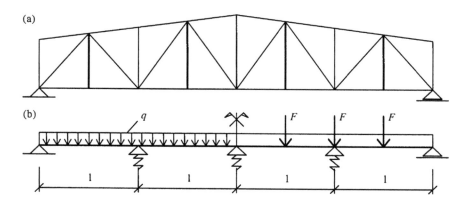

Figure 4.17 (a) Unloading of the lower chord by additional vertical posts, (b) calculation model for the lower chord

By means of additional vertical posts (Figure 4.17(a)), the additional local bending moments M_{loc} on the lower chord at uniform continuous loading q or concentrated force F can be reduced.

$$M_{loc} \cong \frac{1}{16} \cdot q\lambda^2, \; resp. \qquad M_{loc} \cong \frac{3}{16} \cdot F\lambda \qquad (4.2)$$

4.4.2 Change of the static system

The capacity of simple girders can be increased by means of additional elements, while keeping the girder height the same. For example reinforcement can be carried out by introducing new additional, sub-shifted supporting girders between the existing girders (Figure 4.18).

Such adaptation increases the capacity of the floor (or ceiling) and reduces the loading of original girders. This method of reinforcement is effective, but it requires much material. When simple girders are changed into continuous ones (Figure

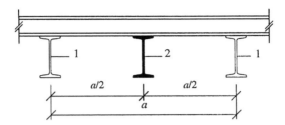

Figure 4.18 Reinforcement by means of additional girders: 1 original girders (supports),
2 added supplementary sub-shifted girder

4.(19) their span is reduced by providing additional intermediate column supports and bearers (Figure 4.19(a)), bearers (Figure 4.19(b)) or a column (Figure 4.19(c)). Providing additional supports (Figure 4.20) changes the magnitude and marks of the bending moment as well as the magnitude of normal stresses. The girder, especially its wall member, should be assessed at the location of the new supports and be modified appropriately.

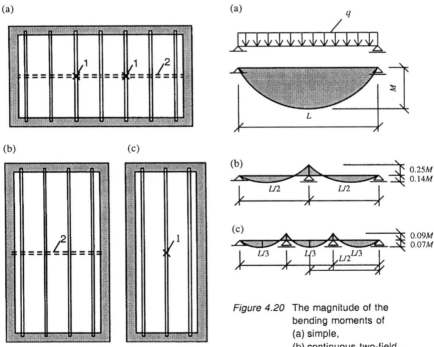

Figure 4.19 Changing simple girders
to continuous ones:
1 column, 2 bearer

Figure 4.20 The magnitude of the
bending moments of
(a) simple,
(b) continuous two-field,
(c) continuous three-field
girders

The gradual change in the magnitude of bending moments (Figure 4.20) shows that in a two-field continuous girder on firm supports, the maximum value for the bending moment can be only 25 per cent of the value for a single girder undergoing continuous loading. However, in the vicinity of the new support, the originally strained chord is changed into compression one, which requires either assessment of the effect of tilting, or a constructional adaptation to stabilize it against buckling. The capacity of single girders can be increased considerably by additional suspension provided by inclined suspenders (Figure 4.21). The method shown in Figure 4.21(a) is suitable for reinforcing roof trusses where the oblique suspenders can be anchored in the upper chords of the trusses that are parts of transverse skylights. The reinforcement in Figure 4.21(b) is less economical because it requires pylons and anchorages in the adjacent fields.

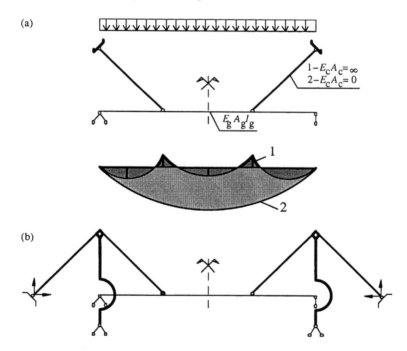

Figure 4.21   Reinforcement by changing the static system using additional inclined suspenders

When there is enough space beneath a girder, it can be reinforced by means of a supplementary draw bar and spreader bars or by creating a strut system (Figure 4.22). In tall struts the theoretical height of the strut is $h\,(1/5 \sim 1/3)\,L$. The draw bar runs a considerable distance from the girder axis; if the draw bar has a suitable geometrical shape and is sufficiently prestressed, the girder is stressed by an axial force at low values of bending moments.

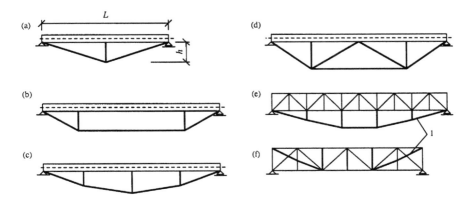

Figure 4.22 Strut systems (a) – (d) used in reinforcing web girders
(e), (f) reinforcement of lattice girders by a third chord, 1 reinforcing third chord

In low struts where the height h does not greatly exceeds the heights of simple girders (1/10 – 1/8) an L-girder is particularly stressed by bending. Figure 4.22 (e), (f) shows the reinforcement of lattice girders by a third chord. The polygonally cranked draw bar requires a special constructional adaptation at the location of three draw bar cranks, where it changes its direction. Where there is not enough space under the girder that is being reinforced (Figure 4.22(f)), it is difficult to create bolsters at the locations of the cranks when placing a draw bar in the frame of the contour of a lattice girder. Sometimes it is better to locate such draw bars along both sides of the girder.

By placing a draw bar outside the girder contours, the magnitude of the bending moments can be reduced considerably. This involves using a direct draw bar (parallel to the girder axis), at some distance from the lower table of the girder, which is anchored in the consoles. The greater the distance the draw bar is from the girder axis, the more economical these struts are, but on the other hand some problems with the stability of anchored consoles can arise. It should also be born in mind that when the draw bar is placed outside the girder contours, the height of the reinforced girder increases.

When frame structures are being reinforced, hinged placements of tie beams can be changed into the solid ones. The result of this change to the static system is a rearrangement of the tie beam and column. When hinges are removed, a three-hinged frame can be changed into a two-hinged or non-articulated one. Figure 4.23 shows the development of moments in changing the static system of the frame. The values for the internal forces of three-hinged, two-hinged or non-articulated frames are given in Table 4.1; in the given case $L = h$ and $I_p / I_s = 1$.

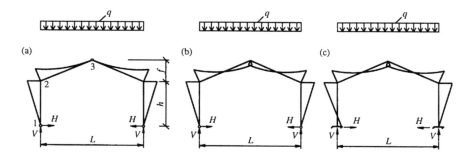

Figure 4.23 Influence of changing the static system of the frame structure on the magnitude and development of bending moments

Table 4.1   The influence of the static system of the frame on the values of internal forces

Internal forces and moments		Frame		
		3-hinged	2-hinged	fixed
$L = h$	H	0.111 g L	0.051 g L	0.087 g L
	M_1	–	–	0.031 g L^2
	M_2	0.111 g L^2	–0.051 g L^2	–0.056 g L^2
	M_3	–	0.068 g L^2	0.058 g L^2

4.5   Reinforcement by additional prestress

To reinforcing structures of girder, frame or arch types can be used. In reinforcement by prestressing, additional stresses of a certain magnitude, direction or duration time are included intentionally built into an element, a part or a whole structure in order to rearrange the internal forces and stresses in the elements of the system. These additional stresses are imposed by mechanical, technological or other means. Prestress can be generated in various ways: by a draw bar made of a high-strength material, by the erection procedure, by temporary change to the static scheme, by changing the position of supports, by inserting some contracting or expanding elements, and by additional loading of overhanging girder ends or frames.

The choice of the appropriate type of prestressing during reinforcement depends on the static scheme, the shape and constructional solution adopted for the load-bearing system, the type and magnitude of loading, and economic and technical consideration. With respect to reinforcement, the use of high-strength prestressing draw bars for imposing the intended shifts or slew is the most effective.

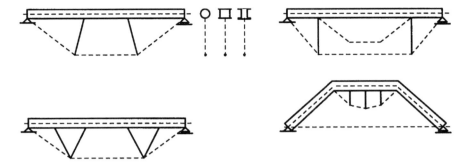

Figure 4.24 Web structures prestressed by draw bars

4.5.1 Reinforcement using prestressing draw bars

Girder structures can be reinforced by prestressing using a direct or cranked draw bar. Some types of web structures prestressed by draw bars are shown in Figure 4.25 (see also Figure 4.22).

Depending on the position, length and shape of the prestressing draw bar, various basic constructional types of web girders can be distinguished (Figure 4.25):
* Prestress by means of a short direct draw bar located within the frame of a girder (close under the lower or upper table and between tables). Only in the most stressed sections in limited parts of the girder is high stress in the girder reduced (Figure 4.25; schemas 7, 10, 12 and 31, 34, 39).
* Prestressing by direct draw bar located within the girder contour. A number of parts of the girder are prestressed, so an efficient redistribution of bending moments in the whole girder is obtained (Figure 4.25; schemas 1, 2, 3, 4).
* Prestressing by cranked or smoothly curved draw bars located within the girder contour. This significantly effects the action of bending moments and of transverse forces in most of the girder (Figure 4.25; schemas 14 – 17, or 43 – 47). It is useful to prestress cranked and curved draw bars simultaneously at both ends of the bars, given the friction in the draw bar cranks. Web structures of the girder type can have additional draw bars located outside the girder contour. The shape of the prestressing draw bar should mirror the stage of the bending moment (e.g. Figure 4.25; schemas 19 to 25).

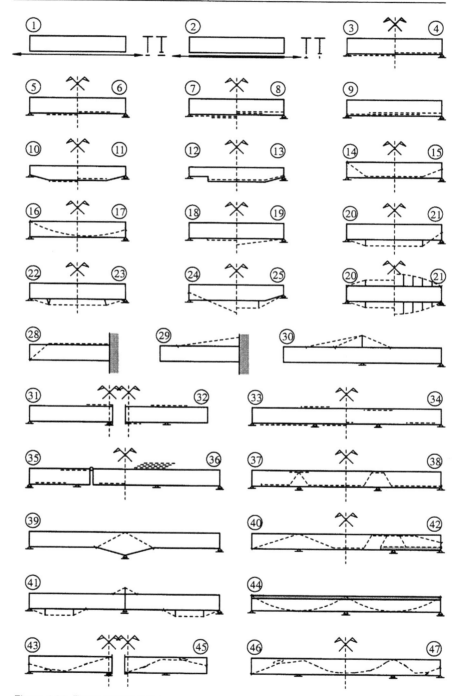

Figure 4.25 Types of web girder

For reinforcing lattice girders, the following constructional types can be distinguished according to the position of the prestressing draw bars (Figure 4.26):

- Where individual members of the lattice girder are prestressed independently (Figure 4.26; e.g. schemas 1 and 3)
- Where several members of the lattice structure are prestressed simultaneously by one or more draw bars located within the structure contour (Figure 4.26; schemas 2, 4, 5; 28 – 32, 7 – 12; 33 – 36). In these schemas the stress on some members may be diminished by prestressing.
- Where the stress on or a majority or all of the members of lattice system is affected by the prestressing draw bar. The draw bar is located completely or partly outside the structure contour (Figure 4.26, schemas 13 – 27 and 37).

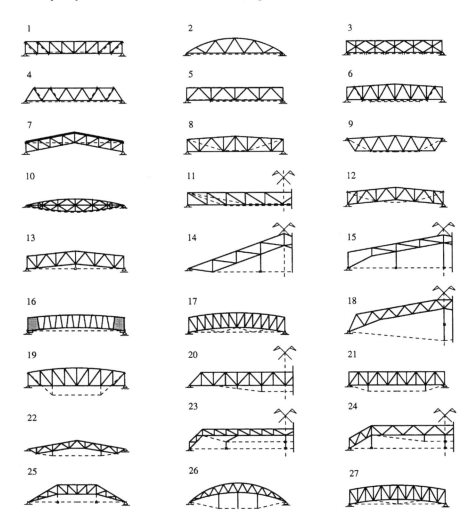

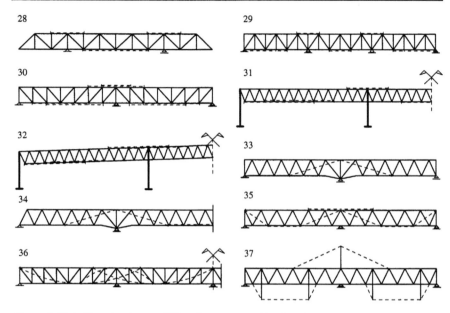

Figure 4.26 Lattice structures additionally prestressed by draw bars

Prestressing by draw bars can also reinforce frame systems with lattice crossbeams (Figure 4.27).

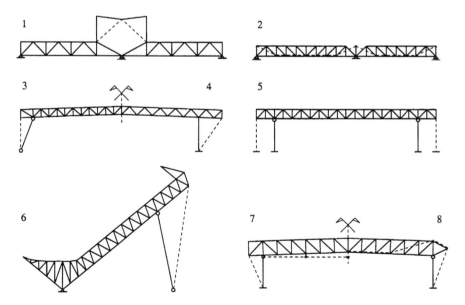

Figure 4.27 Systems reinforced by prestressed draw bars

4.5.2   *Reinforcement by forced deformation*

Prestressing during reinforcing can also be used to redistribute internal forces and stresses by forced deformation of bonds (Figure 4.28). The purpose of the prestressing is to reduce internal forces in the most stressed locations or cross-sections of the structure to the detriment of their increase in the elements or cross-sections that have not been used up till now. In frame and arch structures not only are vertical shifts possible, as in continuous girders (Figure 4.28), but also horizontal shifts on the supports (Figure 4.28(a), (b)). In the case of non-articulated columns ends prestress can be induced by slewing. Other ways of inducing prestress in frame structures, such as by initial eccentricity, additional draw bars and additional loading with ballast, are shown in Figure 4.28(c) – (f).

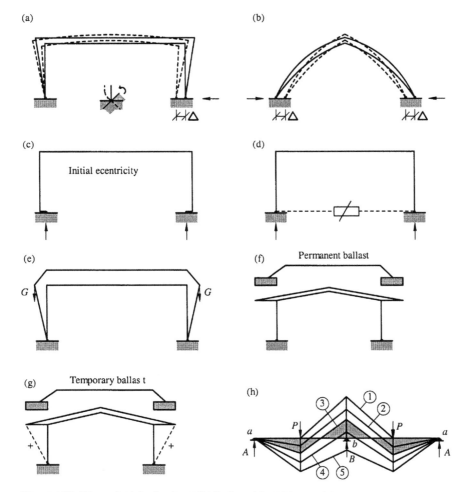

Figure 4.28 Ways of reinforcing by redistribution of forces by prestress

4.6   Reinforcement of steel elements by coupling with concrete

4.6.1   Introduction

Composite steel concrete cross-sections are made by reinforcing the cross-sections of steel elements by coupling them with a concrete reinforcing member. By this reinforcement method cross-sectional resistance is increased, as well as the rigidity of the whole structure, with the aim of increasing resistance against the loss of stability in the shape of an element or of the structure as a whole. At the same time, it is also necessary to pay greater attention to the additional loading of the original element or structure, which is a consequence of this method of reinforcement. Reinforcement by coupling becomes optimal when the concrete member was a part of the original structure but was not acting in conjunction with the steel element. In such a case the dimensions of the steel element of the structure reflect the influence of the dead-weight of the concrete member and their mutual linking into a composite system will be most effective. However, if the concrete reinforcing element is added after the structure has been in use, so that the effects of the dead-weight of the concrete members must be considered sub-sequently, this method of reinforcement is not very effective. Due to low tensional strength, reinforcing coupling concrete members are always located in compressional parts of the cross-section that is being reinforced, when concrete of sufficient burst strength can be used. Inclined elements, or elements stressed by pressure in particular reinforced by coupling.

The concrete reinforcing element can be a single slab or precast. A slab will require a framework and a certain period of time to gain the required strength, which extends both the reinforcing process and down-time of the whole structure or the particular part. This means that during time the steel element that is being reinforced must carry its own dead-weight loading, the weight of concrete mixture and also that of framework (see Figure 4.29(a)). After the concrete part has hardened, the remaining loads are carried by the coupled cross-section (Figure 4.29(b)). To increase the effectiveness of reinforcement by coupling, it is possible to support the given element by erecting scaffolding during concreting and concrete hardening. This makes it possible to take advantage of the linked cross-section for all loads that affect an element or the structure as a whole (Figure 4.29(c)). However, in principle the lower fibres of the original steel cross-section will determine by their recoil the reliability of the cross-section reinforced by such linking.

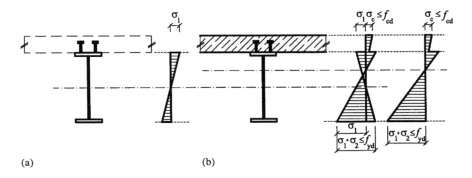

Figure 4.29 Development of stresses along the cross-section stressed by bending, reinforced by linking with the concrete part

When a single monolithic slab is used with steel cross-sections of the third or fourth class, it is necessary to take into account the effect of concrete creeping and shrinkage. These deformational processes are a function of time; at the linkage to the steel elements, changes occur in the pattern of stress along the linked cross-section. Over time the stress moves into the steel element, which will then be stressed more than it was when the cross-section was brought into operation, when the cross-sectional reinforcing procedure and the deformational phenomena are considered.

In the case of cross-sections of the first and second class it is possible to use the plastic resistance of the reinforced coupled cross-section (see Figure 4.30) with respect to the limit capacity state. Here is not necessary to consider either the deformational phenomena or the reinforcing method. However, in accord with the relevant standards [40], [41] it is necessary to check the limit state of usability by demonstrating that during operation the element will be in a flexible state.

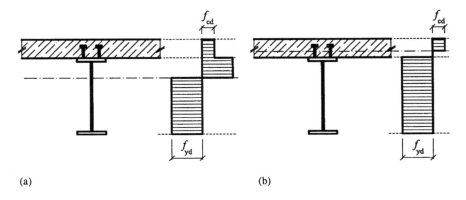

Figure 4.30 Plastic resistance of the cross-section stressed by bending

In this case the normal stress in the cross-section is treated as in cross-sections of the third or fourth class, with respect to the reinforcement method and deformational phenomena. In this method of reinforcement, concrete of the usual composition from C16/20 to C35/45 is used. There is no need to use a high-quality concrete because in most cases when the reliability is being assessed, the steel cross-section that is being reinforced is more important. As coupling elements, plugs or Hilti stops are particularly useful because of their quick mounting by cartridge hammer fixing onto the upper surfaces of the steel elements. For plate steel elements, coupling bars that act as longitudinal stiffeners supporting the plate at the stage of concreting can be used.

Some deficiencies in this method of reinforcement by coupling with monolithical concrete slabs can be removed by using precast units. Precast concrete reinforcing elements are most frequently coupled by means of prestressed high-strength bolts (Figure 4.31).

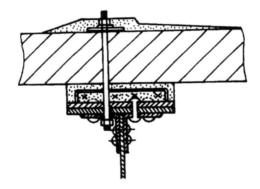

Figure 4.31   Reinforcement by coupling using high-strength bolts

When checking the reliability of a cross-section reinforced by coupling with precast concrete units, the same procedure as for monolithical concrete can be used. In flexible checking of the reliability of a reinforced cross-section, it is possible to ignore the deformational influences, which are not important in the case of precast elements.

4.6.2   Examples of reinforcement of steel elements by coupling with concrete

Bridge are classical examples of reinforcement by coupling with the concrete. Land communication bridges with flooring require a load-bearing base for the deck, which is usually a reinforced concrete plate. This is set on the flange plate of the ledgers and binders, but is not linked to them. The lifetime of this method is shorter in comparison with that of the steel elements of the load-bearing structure, so it must be replaced during the lifetime of the bridge. During this renovation it is possible to attain a higher resistance in the flooring elements by coupling them with a newly made steel concrete load-bearing base.

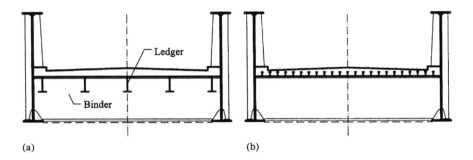

(a)   (b)

Figure 4.32 Reinforcement of flooring by coupling with a reinforced-concrete load-bearing sub-base

As shown in the example in Figure 4.32, there is an advantage, where if possible given the geometric parameters of the bridge structure and sufficient inclined resistance of the binders, to eliminate ledgers and couple the reinforced concrete plate with binders only. The weight of the flooring is thereby reduced and the structure is more efficient. If this is not possible, the plate can be additionally coupled in the longitudinal direction. The thickness of the reinforced concrete plate depends on the type of linking (longitudinal or transversal), which determines the computing model for the plate. The plate is usually 200 mm thick, reinforced by the normal concrete reinforcement. This type of reinforcement is used less often in the flooring of railway bridges. In the case of railway bridges the reinforced concrete plate is incorporated in the bridge system together with the continuous track bed. However, this considerably increases the weight of the flooring, for which the main load-bearing system has usually not been designed. In addition, incorporating a reinforced concrete plate in the lower tensional elements of the bridge brings complications caused by significant interaction of the reinforced flooring with the main girders, which are manifested in tensional stress in the reinforced concrete plate. In such cases it is necessary to pay close attention to the limit states for crack occurrence and widths and especially important to use a very high-quality and flexible water-separation material under the railway bed. Although this method of reinforcing the flooring is not optimum, there are some examples of its use. It was used in reinforcing the flooring and making the continuous track bed on the railway bridge over a river on the Bologna – Verona route in Italy [42]. The cross-section of the bridge together with the newly built-in reinforced concrete plate with continuous track bed is shown in Figure 4.33.

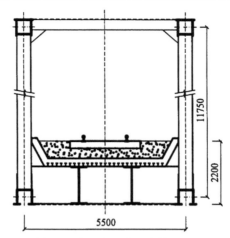

Figure 4.33 Reinforcement of the flooring of a railway bridge by coupling with a reinforced-concrete plate with the continuous track bed

The reinforcing used a 300 mm thick plate linked in both transverse and longitudinal directions with binders as well as with ledgers using plugs. By increasing the constructional height of the bridge by about 600 mm, a high quality carriage way, which increased comfort of passengers and reduced noise. Beside the improvement in operational characteristics, strengthened durability of the bridge structure was increased considerably, as of course was its weight. It was necessary to take great care with water separation which was effected by a 4 mm thick synthetic polyurethane coating. Except that the anticorrosive stainless reinforcement was applied.

The reinforcement of the main load-bearing system of a bridge by coupling with a reinforced concrete plate is important only for bridges with upper flooring

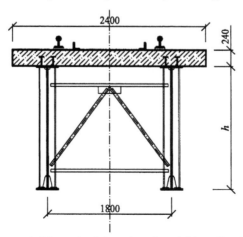

Figure 4.34 Reinforcement of the main girders of a railway bridge without flooring by coupling with a reinforced-concrete plate

or without flooring together. In such cases, the reinforced concrete plate replaces the original bridge superstructure. A possible application of such reinforcement is shown in Figure 4.34.

Where there is sufficient load-bearing capacity of the main girders, or in combination with reinforcement of lower chords, it is possible to reinforce the main girders with reinforced concrete plate that at the same time form the channel for continuous track bed. An application of this solution is given in Figure 4.35.

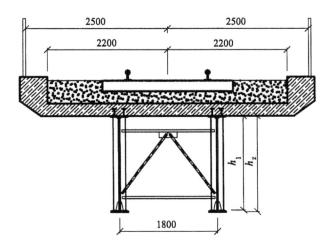

Figure 4.35 Reinforcement of the main girders of a railway bridge without flooring by coupling with the channel of a continuous track bed

From the authors' own experience, we can give an example of reinforcement of a load-bearing structure by coupling with a reinforced concrete monolithic plate [43]. The case concerned the load-bearing structure for the equipment used to catch and remove dirt from water at the works at Hrièov. The original load-bearing structure was formed by a steel chamber through-beam split into two branches with field spans of 17 + 22 + 20 m in one branch and 21 + 21 m in the other, with the axes of the branches forming 90° angle. The chamber beam is 1500 mm high and 2000 mm wide. Because of the narrow thickness of the chords and walls (8 −10 mm), it was necessary to reinforce them. The upper chord was reinforced by nine longitudinal reinforcements of I 100 section welded by the upper flanges to the chord plate. The lower chord was reinforced by just one longitudinal reinforcement of I 100 section, while the walls were reinforced by 100 × 100 × 8 mm square section. At each 2 m internal transverse reinforcements of chords and walls were inserted. On the lower transverse reinforcement, the rack to catch the dirt was hung by means of hinges. The structure is in effect a bridge that carries the equipment for catching dirt before water flows out from the dam, and at the same time it serves as a bridge for a light vehicle with a clamshell, by which dirt is picked out

Figure 4.36 View of a load-bearing structure for equipment for catching and removing dirt

and removed to the stock pile. The photograph in Figure 4.36 and a cross-section of the structure in Figure 4.37 may serve to illustrate this description.

The need to increase the weight of the vehicles that pick out and remove the dirt resulted in the need to check the load-bearing system. Static calculations showed a low loading capacity of the orthotropic flooring, which was insufficient to accommodate the movement of the wheeled vehicle required.

Because the main load-bearing system had a sufficient capacity reserve, the suggested solution was to reinforce the orthotropic flooring by coupling it with a monolithic reinforced concrete plate which was laid on the existing flooring to an

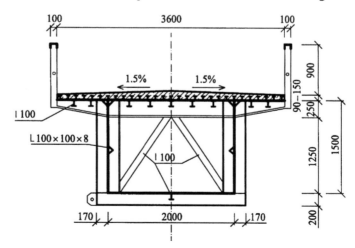

Figure 4.37 The cross-section of a load-bearing structure after reinforcement of the flooring by coupling with a reinforced-concrete plate

average thickness of 120 mm (see Figure 4.37). The coupling was secured by plugs, alternating with perforated coupling bars that increased the strength of the chord during the concrete hardening stages.

In some land constructions, cases of reinforcement by coupling with a concrete part will be found used in the refurbishment of load-bearing ceiling structures. The coupling of a new monolithic ceiling plate with the original steel ceiling beam is an effective way of increasing the capacity of a ceiling structure.

4.7 Reinforcement of elements under loading

4.7.1 Introduction

Steel is a suitable material for refurbishing elements of steel structures by reinforcement, with the objective of enlarging the sectional characteristics of the existing cross-section. The reinforcement is often carried out on the elements directly in the steel structure, i.e. without removing them out from the system, because this method is difficult and requires either a shut-down of the structure or secondary support in the affected positions in the structure. Reinforcement of elements under loading is by comparison less effective, to be used only for a short-term loading that can be dealt with during refurbishment by restricted operation or short-term stoppages, or if it involves a climatic loading (snow, wind) by avoiding times when such loading is active during the refurbishment. Reinforcement of elements of steel structures under loading is a difficult problem from both theoretical and practical points of view. In principle it is always more difficult than it is to design and fit of a new element.

4.7.2 Reinforcement of axially stressed elements

Elements stressed principally by axial forces are most often found in the lattice structures of load-bearing roof structures or in load-bearing bridge structures.

Tensional elements

In tensional members reinforcement of their cross-sections is relatively simple. Here the cross-sectional stress is important, so the relevant condition is not to exceed the design value for the resistance of the stressed (original) cross-section that is being reinforced. In bridge structures, drawing on the plastic reserve of steel [37], [39] is normally not recommended, so in checking the reliability of the reinforced cross-section the relevant condition is defined as:

$$\frac{N_{Sd,1}}{A_1 f_{yd}} + \frac{N_{Sd,2}}{(A_1 + A_2) f_{yd}} \leq 1 \tag{4.3}$$

where

A_1 is the area of the (original) cross-section of the element that is being reinforced

A_2 is the area of reinforcement of the cross-section

$N_{Sd,1}$ is the design value of the axial force in the original cross-section of the element during reinforcing (usually the axial force from the permanent loading)

$N_{Sd,2}$ is the design value of the additional axial force loading the cross-section after reinforcement, i.e. the axial force from the short-term loading that was then eliminated during reinforcement

$f_{yd} = f_y / \gamma_{M0}$ is the design value of the material resistance of the reinforced cross-section.

From equation (4.3), after rearrangement we obtain the necessary area of reinforcement A_2:

$$A_2 \geq A_1 \left(\frac{N_{Sd,2}}{f_{yd} A_1 - N_{Sd,1}} - 1 \right) \tag{4.4}$$

The idea of this method of reinforcing the cross-section of a tensional element is illustrated in Figure 4.38a.

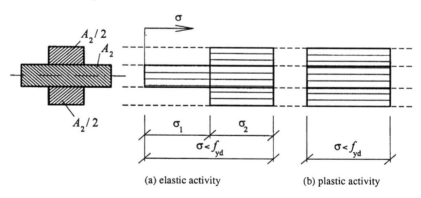

(a) elastic activity   (b) plastic activity

Figure 4.38 State of stress of the reinforced cross-section of a tensional element

The procedure assumes a flexible assessment of the cross-section of a tensional element. For land constructions, this procedure is conservative because we can use plastic condition in checking the reliability of the cross-section of a tensional member in shapes that assume compensation of the stress over the whole cross-section, i.e. the full use of reinforcing elements in the cross-section, as shown in Figure 4.38b:

$$\frac{N_{Sd,1} + N_{Sd,2}}{(A_1 + A_{2p})f_{yd}} \leq 1.0 \tag{4.5}$$

The necessary area A_2 for transferring the loading is determined by equation (4.5) to give:

$$A_{2p} \geq A_1\left(\frac{N_{Sd,1} + N_{Sd,2}}{A_1 f_{yd}} - 1\right) \tag{4.6}$$

To distinguish equations (4.5) and (4.6), the area of reinforcement is stated as A_{2p}. By comparing equations (4.6) and (4.4), we can determine the efficiency of the plastic condition approach.

$$A_{2p} = A_2\left(1 - \frac{N_{Sd,1}}{A_1 f_{yd}}\right) \tag{4.7}$$

From equation (4.7) it is apparent that the use of the plastic reserve of steel is more significant and the saving of steel is greater the higher the original stress on the reinforced cross-section.

Compressional elements
The reinforcement of compressional elements involves complicated cases of reinforcing. By contrast with tensional elements, here it is important to locate correctly the reinforced elements of the cross-section. In reinforcing the cross-section of a compressional element, not only this cross-sectional area but also the strength of the complete element is altered, which affects the width – thickness ratio of the member. By suitable placement of the reinforcing element, it is possible to save an amount of filler material. From the theoretical point of view the reinforcement of compressional element is a complicated problem because given the techniques of connecting the filler material, the imperfections of the reinforced element and thereby also the overall behaviour of the reinforced element, are both changed.

By application of modern concept of the buckling of compression members [38] it is possible to determine the necessary reinforcing area of the most-stressed cross-section of a compression element by checking the reliability of the most-stressed cross-section:

$$\frac{N_{Sd,1}}{\chi_1 A_1 f_{yd}} + \frac{N_{Sd,2}}{\chi_2(A_1 + A_2)f_{yd}} \leq 1.0 \tag{4.8}$$

where A_1, A_2, $N_{Sd,2}$ are defined by equation (4.3)

$f_{yd} = f_y/\gamma_{M1}$ is the design value of the material resistance of the reinforced
cross-section

χ_1 is the buckling coefficient of the original element before reinforcement
χ_2 is the buckling coefficient of the reinforced element.

From the reliability condition given in equation (4.8), the necessary area of reinforcement A_2 is given by the equation

$$A_2 \geq A_1 \left[\frac{\alpha_2}{\xi(1-\alpha_1)} - 1 \right]$$ (4.9a)

where

$$\alpha_1 = \frac{N_{Sd,1}}{\chi_1 A_1 f_{yd}} \qquad \alpha_2 = \frac{N_{Sd,2}}{\chi_1 A_1 f_{yd}}$$ (4.10)

$$\xi = \chi_2/\chi_1$$ (4.11)

From equation (4.9a) it follows that the necessary area of reinforcement of the cross-section of compression element is determined only on the condition that we know the value of buckling coefficient after reinforcement χ_2. It is possible to apply an iterative procedure by choosing $\xi = \chi_2/\chi_1 = 1.0$ for the first interaction. After determining the required area A_2 of reinforcement of the cross-section, the value of buckling coefficient χ_2 is calculated according to the position of the filler reinforcing material, and subsequently we can improve the calculation of the required area A_2. However, we can also find the optimum necessary area of the reinforcement by calculations performed given the condition that the requirement of efficiency of the reinforcement is met. This requirement is given by

$$\left(N_{b,Rd,2} - N_{b,Rd,1}\right)/\left(N_{c,Rd,2} - N_{c,Rd,1}\right) > 1.0$$ (4.12)

Equation (4.12) expresses the condition that the increase in resistance of the element after reinforcement must be greater than the increase in the simple pressure resistance. This condition at the same time takes into consideration the optimal position of the reinforcing filler material in the reinforced cross-section, in the sense that the same value of the buckling coefficient is at least maintained after reinforcement. The condition for optimum use of added material leads after rearranging to the equation expressing the optimum rate of buckling coefficients after and before reinforcement:

$$\xi \geq \frac{1+\varphi}{\chi_1(1+\varphi)}$$ (4.13)

where $\xi = \chi_2/\chi_1$ and $\varphi = A_2/A_1$ (4.14)

If we substitute equation (4.9a) in modified form

$$\varphi = A_2 / A_1 = \frac{\alpha_2}{\xi(1-\alpha_1)} - 1$$ (4.9b)

in the equation (4.13), we obtain an equation for optimum value of the buckling coefficient χ_2 of the element after reinforcement:

$$\chi_2 \geq \frac{\alpha_2 \chi_1}{1-\alpha_1 + \chi_1(\alpha_1 + \alpha_2 - 1)}$$ (4.15)

By inserting values in equation (4.5) and then substituting the result in equation (4.9a), we obtain the optimum area of reinforcement of the cross-section of the compressional element. This procedure ensures optimum use of reinforcing material. However, in practice we may not always, manage to obtain such the optimum state, because it depends on the position of filler material and it is not always possible to place it in the optimum position.

As an example of the reinforcement of the cross-section of a compressional element, there is possibility of adding the reinforcing material in rolled I 200 section, where the deflection in the direction of the z axis is crucial. Before reinforcing the section, the relevant cross-section of the compressional chord has these parameters:

$A = 3.35 \times 10^{-3}\,\text{m}^2$
$i_y = 0.08\,\text{m}$
$\chi_1 = 0.693$
$f_{yd} = f_y/g_{M1} = 235/1.10 = 213.64\,\text{MPa}$

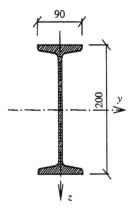

Figure 4.39 Cross-section of a compression member

It is necessary to reinforce the cross-section so that it can accommodate the increase in axial force given by the degree (rate) of additional loading:

$$\alpha_2 = \frac{N_{Sd,2}}{\chi_1 A_1 f_{yd}} = 0.706$$

From equation (4.15) we determine the optimum buckling coefficient after reinforcement, which is $\chi_2 0.763$ after substitution into the above quantities that corresponds to a ratio $\xi = 1.10$. From equation (4.9a) we calculate the necessary area of reinforcement of the original I 200 cross-section for the values of $A_2 = 0.294$ and $A_1 = 0.985 \times 10^{-3}$ m². If this area is provided by stiffening plates 2 100 × 5, which are welded on the flanges of the original section according to Figure 4.40(a), the reinforced cross-section will have these parameters:

$A_1 + A_2 = 4.35 \times 10^{-3}$ m²
$i_z - 0.0856$
$\chi_2 = 0.724$

However, note that for the given parameters the reinforced cross-section does not meet the condition given by equation (4.8).

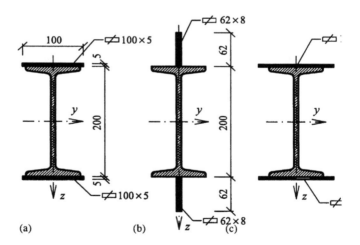

Figure 4.40 Ways of reinforcing the cross-section of a compression member

If we check the reinforcement of the original cross-section given in Figure 4.40 (b), we obtain a considerably more favourable value for the buckling coefficient of $\chi_2 = 0.769$, which meets the optimum condition, and the reinforced cross-section also meets condition (4.8). Figure 4.40(c) shows the optimum case of reinforcement of the original section for buckling in the direction of y axis y using the same loading parameters of members as for buckling in the direction of z axis. This form of reinforcement shows a buckling coefficient value of $\chi_2 = 0.8$, which is higher than needed. The cross-section therefore meets the condition of reliability (4.8).

In the case of cross-sections of the first and second classes stressed by static loading, it is possible to use their plastic reserve and to check the reliability of the cross-section by the condition

$$\frac{N_{Sd,1} + N_{Sd,2}}{\chi_2 (A_1 + A_2) f_{yd}} \leq 1.0 \tag{4.16}$$

As the quality of the steel in the reinforced element cannot always be guaranteed, and because not enough is known about the plastic reserve of older steels, generalising the procedure associated with equation (4.16) is not recommended. A better way of saving on the quantity of reinforcing material used is by searching for its optimum location on the reinforced cross-section of compressional element. In slender cross-sections of the fourth class, the effective areas of cross-section $A_{eff,1}$, respectively $A_{eff,2}$ are used, which are determined by procedures according to the relevant standards [38], [39]. Using suitable methods of reinforcing the cross-section, it is possible to prevent buckling (swelling) of the slender parts of the cross-section.

4.7.3 Elements stressed by bending

The cross-sections of elements subject to bending represent the most frequent cases for reinforcement. To clarify such cases, we consider a symmetrical cross-section of the third class stressed by bending moment $M_{Sd,1}$ which in the outer region of the cross-section applies a normal stress σ_1 (see Figure 4.41). The cross-section should be reinforced so that it can carry the increased bending moment $M_{Sd,2}$.

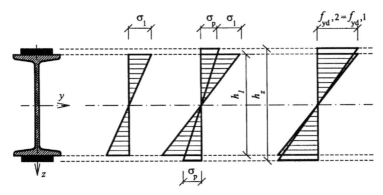

Figure 4.41 The developments of stresses during reinforcing a cross-section of the third class

The outer region of the original reinforced cross-section must meet the condition of reliability given by

$$\frac{M_{Sd,1}}{W_{el,1} f_{yd,1}} + \frac{M_{Sd,2}}{W_{el,z} f_{yd,1}} \frac{h_1}{h_z} \le 1.0 \qquad (4.17)$$

where

h_1 is the height of the original cross-section before reinforcement
h_2 is the height of the reinforced cross-section
$W_{el,1}$ is the modulus of elasticity before reinforcement
$W_{el,z}$ is the modulus of elasticity after reinforcement
$f_{yd,1} = f_{yl}/\gamma_{M0}$ is the design value of the material resistance of the reinforced cross-section. If we reinforce the cross-section with material of the same quality as the material in the original cross-section, reliability condition (4.17) ensures the reliability of the whole cross-section, including the reinforcing elements. If the quality of the materials is different, it is necessary also to check the outer region of the reinforcing elements using the condition

$$\frac{M_{Sd,2}}{W_{el,z} f_{yd,2}} \le 1.0 \qquad (4.18)$$

We obtain the optimum reinforcement of a cross-section stressed by bending if the outer regions of both original and reinforced cross-sections are equally used (see Figure 4.41). This condition means that the design resistance value is reached simultaneously in both outer regions. Applying this on condition that $f_{yd,1} = f_{yd,2} = f_{yd}$, we obtain the value of the required ratio of cross-section heights (h_1) before and (h_2) after reinforcement:

$$\frac{h_1}{h_z} = 1 - \frac{M_{Sd,1}}{W_{el,1} f_{yd}} \qquad (4.19)$$

However, in practice we hardly ever obtain the optimum height ratio in equation (4.19) because equation (4.19) requires an excessive enlargement of the cross-section that technically is not possible to attain. We usually propose the reinforcement so that the height ratio before and after reinforcement is greater than that given in equation (4.19) so that the stress in the reinforced part is decisive. In the less frequent case when the height ratio is lower than that required by equation (4.19), the reinforcing part limits the reliability.

In case of cross-sections of first and second classes, it is possible to use their plastic reserve by partial or full plastification of the cross-section. We can reach the partial plastification of the original cross-section if the condition of reliability chosen is just equation (4.18), which controls the elastic state in the outer regions of the reinforcing part of cross-section but ignores tensions in the outer regions of the reinforced part.

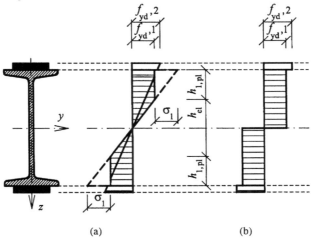

(a)   (b)

Figure 4.42 Reinforcement of a cross-section using the plastic reserve of the material

From the condition for obtaining the design resistance value $f_{yd,1}$ of the original material in a region a distance $h_{el/2}$ from the neutral axis of the cross-section (see Figure 4.42(a) given by

$$\frac{M_{Sd,1}}{W_{el,1}}\frac{h_{el}}{h_1}+\frac{M_{Sd,2}}{W_{el,z}}\frac{h_{el}}{h_z}=f_{yd,1} \tag{4.20}$$

we determine the height of the elastic part of the cross-section:

$$h_{el}=\frac{h_1}{\dfrac{f_{yd,2}}{f_{yd,1}}\dfrac{h_1}{h_z}+\dfrac{M_{Sd,1}}{W_{el,1}f_{yd,1}}} \tag{4.21}$$

from which the height of the plasicized part of the cross-section is given by

$$h_{1pl}=\frac{h_1-h_{el}}{2}=1-\frac{1}{\dfrac{f_{yd,2}}{f_{yd,1}}\dfrac{h_1}{h_z}+\dfrac{M_{Sd,1}}{W_{el,1}f_{yd,1}}} \tag{4.22}.$$

In complete plasticizing, for the original as well as the added reinforcing material, their bending resistance is determined by use of the tension developed according to Figure 4.42(b):

$$M_{pl,Rd} = f_{yd,2}\left(W_{pl,z} - W_{pl,1}(1 - \frac{f_{yd,1}}{f_{yd,2}})\right)$$

(4.23)

where

$W_{pl,z}$ is the modulus of elasticity of the reinforced cross-section
$W_{pl,1}$ is the modulus of elasticity of the original cross-section being reinforced.

The reliability of the cross-section is verified if condition (4.24) is met:

$$M_{Sd,1} + M_{Sd,2} \leq M_{pl,Rd}$$

(4.24)

From the reliability condition (4.24) it is possible to determine the area of the cross-section that needs reinforcing so that it can carry the additional loading specified by the bending moment $M_{Sd,2}$. The total area of reinforcement required is determined by the relation

$$A_2 \geq 4\left(\frac{M_{Sd,1} + M_{Sd,2} - W_{pl,1}\,f_{yd,1}}{h_1(1+\psi)f_{yd,2}}\right)$$

(4.25a)

where $\psi = \dfrac{h_z}{h_1}$ is the height ratio of the cross-section before and after reinforcement.

If the additional loading and growth in bending moment $M_{Sd,2}$ is not significant, it is possible to consider $\psi \approx 1$ and relation (4.25a) becomes simpler. This enables direct calculation of the required area of reinforcement:

$$A_2 \geq 2\left(\frac{M_{Sd,1} + M_{Sd,2} - W_{pl,1}\,f_{yd,1}}{h_1\,f_{yd,2}}\right)$$

(4.25b).

The case of reinforcement of an asymmetric cross-section is a generalization of the procedure for reinforcing the cross-section stressed by bending. If we consider the limit state of the reinforced cross-section to be attaining the design resistance of the material in the outer region of the reinforced original cross-section, condition (4.17) can be generalized. For the upper region,

$$\frac{M_{sd,1}}{W_{el,11}f_{yd,1}} + \frac{M_{sd,2}}{W_{el,z1}f_{yd,1}} \frac{e_1 + \Delta e}{e_{z1}} \le 1.0 \tag{4.26a}$$

and respectively for lower fibres:

$$\frac{M_{sd,1}}{W_{el,12}f_{yd,1}} + \frac{M_{sd,2}}{W_{el,z2}f_{yd,1}} \frac{e_2 - \Delta e}{e_{z2}} \le 1.0 \tag{4.26b}$$

where

$W_{el,11}$ is the modulus of elasticity of the cross-section for the upper region of the original cross-section

$W_{el,12}$ is the modulus of elasticity of the cross-section for the lower region of the original cross-section

$W_{el,z1}$ is the modulus of elasticity of the cross-section for the upper region of the reinforced cross-section.

$W_{el,z2}$ is the modulus of elasticity of the cross-section for the lower region of the reinforced cross-section.

Other terms are shown in Figure 4.43.

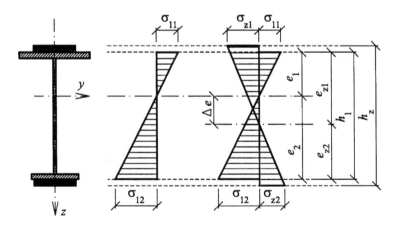

Figure 4.43 Reinforcement of an asymmetric cross-section

Attaining design resistance simultaneously in both the outer regions of the original cross-section can be considered as the condition of economic efficiency of reinforcement. By comparing relations (4.26a) and (4.26b), we obtain a formula for moving the centroidal axis of the reinforced cross-section:

$$\Delta e = \frac{e_2}{h_1}(e_2 - e_1)(1 + \frac{M_{sd,1}}{M_{sd,2}}\zeta)$$ (4.27)

where $\zeta = \frac{I_{yz}}{I_{y1}}$ is the ratio of the moments of inertia of the reinforced and original cross-sections.

This approach is used not only in the case of hemicompact cross-sections of the third class for which the derivation was provided, but is also applicable also for cross-sections of the fourth class. In these cross-sections, the asymmetry originates from buckling of slender parts, particularly walls, which is taken into account in practical calculations by using the effective cross-sectional area according to standards [38], [39]. The bent cross-sections show a shifting of the centroidal axis of the effective cross-section due to buckling of the wall or flange, so the above procedure becomes applicable in cases of symmetrical slender cross-sections. Elastic compact cross-sections can be checked according to the principles of elasticity used in relation (4.25).

4.7.4   Techniques for reinforcing elements under loading

When reinforcing an element directly in a structure, it is advantageous to connect the reinforcing part to the cross-section being reinforced by welding. This requires a good weldability of materials, which can be a problem for old structures from the beginning of twentieth century. The advantage arises because the variability of cross-sectional shape gives more possibilities for the location of the reinforcing part. Welding is particularly suitable in riveted cross-sections where clinching is not necessary. When welding the added part it is necessary to pay close attention to the quality of the welds and to follow the procedure recommended by the welding technologist. The welding procedure should eliminate unfavourable tensions caused by welding that can cause deformation of the element as well as affect their resistance in compression members. In principle, with smaller cross-sections we prefer symmetrical welds realised by electrodes. It is necessary to avoid the accumulation of welds, saw-cuts and the details that lead to points of stress, which can initiate brittle or fatigue fracture in dynamically loaded structures.

Possibilities for reinforcement by welding of some basic cross-sections of elements of steel structures for the usual cases of rolled and welded sections were shown in Figures 4.3 and 4.4. Possibilities for reinforcement of riveted cross-sections by welding the reinforcing part are shown in Figure 4.44.

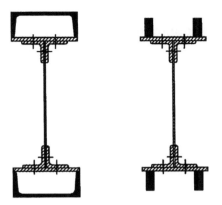

Figure 4.44 Ways of reinforcing a riveted cross-section by welding the reinforcing part

In non-weldable material the reinforcement of a riveted cross-section can be done by adding stiffening plate at suitable angles. However, this requires clinching the cross-section, so its tension will be changed if the reinforcement is realised directly in the structure. If rivets are loosened, the co-operative action of the flanges with the wall is broken, which leads to increased tension in the remaining parts of the structure during refurbishment. The efficiency of reinforcement is therefore lowered considerably because the renovation of the cross-section with the added reinforcing part is ineffective for active tension and is effective only for loading that will be excluded during reinforcing. The rivets in the cross-section being reinforced are usually replaced by bolts; in elements of bridge structures we prefer high-strength 8.8 and 10.9 bolts. Examples of reinforcement of typical riveted cross-sections are shown in Figure 4.45.

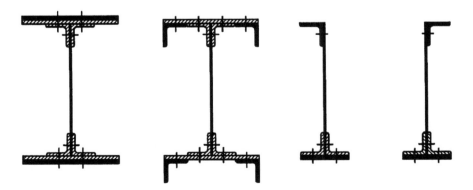

Figure 4.45 Examples of reinforcement of riveted cross-sections

4.8   Finding capacity reserves in original structure by accurate assessment of its actual operation

4.8.1   Introduction

By assessing the actual operation of the original structure more accurately, it may be possible to find some reserve capacity. The extent of reinforcement can be reduced if such reserves can be established and used, which reduces the material and time needed for the refurbishment. The procedure used to find capacity reserves in a structure is shown in Figure 4.46. The capacity reserves of a structure can be determined by an accurate static calculation in which the factors determining the capacity of the structure (loading, static scheme, calculation model, steel strength) are considered in more detail than they were in the original design.

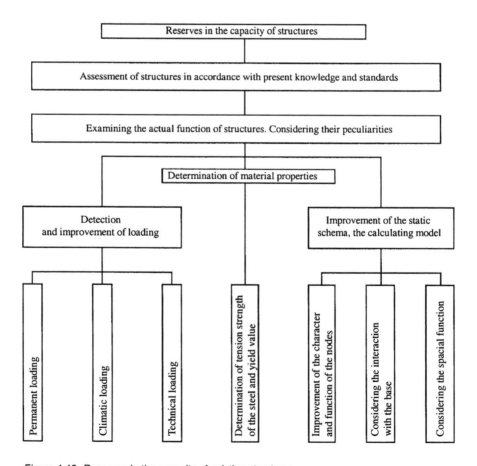

Figure 4.46   Reserves in the capacity of existing structures

Individual characteristics of the refurbished structure, such as local conditions, technology, modes of operation, and peculiarities of the constructional method must be considered.These factors are incorporated in the design values for technical and climatic loadings and thereby also the design values for internal forces in the various elements of the structure. Snow and wind loadings can be determined on basis of the data from the nearest meteorological station.

Existing steel structures were designed in accordance with the procedures and standards valid at the time of designing the structure. In the course of development of the theory and practice of designing structures, methods of design and of checking structure reliability have been improved, with the aim of determining their actual operation as accurately as possible. The development of theory of structural design corresponds closely to the development of computerized techniques, whose introduction into the design process meant a significant qualitative change that has worked backwards to influence design procedures, and forwards to methods of checking structural reliability. The present possibilities of computer techniques and their continuous development are becoming driving forces in the development of the theory of structures and are existing a great influence on the standards being set for structural design. In this connection should be mentioned the grand project of unified European standardisation, which represents the leading edge current approaches to the theory of structural design. It is this perspective on present knowledge in this field that the assessment of existing is discussed here.

It is obvious that such higher precision approaches to the design of steel structures will reveal possible reserves in such structures. These reserves offer a number of new possibilities:

- New methods of assessing structure reliability
- More information on the loading of building structures and new possibilities for assessing their magnitude, especially of permanent (though changeable) long-term loads
- High-quality transformational numerical models capable of determining non-linear activity, in a structure
- Higher-quality and deeper information on structural materials enabling the intentional utilization of the elastic properties of steels.

4.8.2   New procedures for assessing the reliability of steel structures

The majority of existing steel structures were designed according to the method of permissible stresses. This is a deterministic method of assessing the reliability of structures that compares characteristic or nominal effects of loading, usually in stress condition, with the specified permissible stress. A brief description of this method was given in Chapter 1 of this book. The present procedure for checking reliability is based on a semi-probabilistic method of partial coefficients, also referred as to the method of limiting conditions. This method uses design values for the effects of loading and the resistance of a material, an element or a structure. Design values with a probabilistic interpretation can be compared with the prescribed quartiles for their occurrence. In cases where there is no background statistical distribution for these values or it is insufficient, we can empirically assess values determined by means of characteristic values (eventually nominal) and partial coefficients for load and material reliability. This method was also described in Chapter 1.

The method of partial coefficients has been used for designing steel structures under current conditions for almost 40 years and so has attained a very high standard. With respect to probabilistic aspects it is of higher standard than the method of permissible stresses. Because the method of partial coefficients works with statistically processed data, it provides with better information on the loading of structures. For example the data on the strength characteristics of steel have been monitored in Slovakia since the 1960s, providing a great amount of statistical data that makes it possible to specify real design values for the strength characteristics of steels.

Given this, we can expect some reserve capacity in the permanent and variable long-term loading of building structures where there is high-quality data, so design values for these loadings can be determined more accurately. The largest reserves can be expected in structures where the effects of permanent and variable long-term loading are large. Similarly, with respect to monitoring the quality of steel, we can also expect here a higher standard of mechanical properties in comparison with the data used at the time of designing. Other reserves result from modification of the procedures involving partial coefficients for checking the reliability of existing structures. This is much more data on existing structures because it is now possible to measure their actual geometry including, imperfections and defects. This provides more accurate values for permanent and variable long-term loadings, as well as more precise durability characteristics for the materials used, and therefore the initial basic parameters involved in the process of reliability assessment of such structures is reduced. The design values for the effects of loading and resistance of the structure for a lower level of reliability as it is required for new structures when the above described information are unknown. This idea was also explained in more detail in Chapter 1.

From this brief survey, it is apparent that the method of partial coefficients including its modifications for existing structures, represents a more economical method of checking the reliability of building structures, given the possibility of using the reserves resulting from that. Modern approaches to the assessment of existing bridge structures are an example of the application of such considera tions. These ideas are gradually being incorporated in standard regulations and procedures for the assessment of existing bridges [44], [45], [46] and translated into real values for the reliability of existing bridges.

4.8.3   The influence of the precision of values of loading

The basic function of a building structure with respect to its reliability is its ability to carry the loads acting on it its life time. Therefore it is necessary to pay close attention to the way its loading is calculated. If the loading is determined certain assumptions about its magnitude and the period of activity, which are reflected in the corresponding values of the partial coefficients, in the case of existing structures we can check the magnitude of loading directly on the structure. We can therefore asses the real values of permanent and changeable long-term loadings in the simplest way. However, loadings that vary in the short-term are harder to determine and require long-term monitoring.

More accurate values of permanent and variable long-term loads for existing structures are obtained from the actual geometrical dimensions of the elements or parts of a structure, which are determined by diagnostic research. For the determined values of loading it is again possible to consider a lower value for the partial coefficient of reliability of loading. For example the Standard [47] allows the value of partial coefficient of reliability of permanent or variable long-term loading to be reduced by 0.1. Obtaining a more accurate actual loading may again reveal possible reserves in the existing structure, which can then be used in the assessment of structure reliability.

4.8.4   Development of calculating algorithms for steel structures

The rapid advance of computing technology and the general computerization of the process of structural design has resulted in the development of calculating algorithms for the global analysis of building structures. Nowadays the predominant numerical transformations, are based on finite element methods. In the large majority of existing structures, the internal forces involved in their design were determined using elementary methods of analysing the statics of building structures, compared to the present more sophisticated facilities of methods. Manual calculation of internal forces in structures was performed using individual planes, into which the structure was analysed in order to determine the effects of loading resulting from the basic functioning of each part of the structure. The co-operation of these individual parts of the structure was rarely taken into consideration by using supplementary plane patterns. Present trends prefer now the use of complex spatial calculating patterns that reflect the global activity of the structure, including the redistribution of internal forces. It is necessary to app-reciate that the simple calculating patterns were chosen so that they provided a margin of safety. This means that the use of more accurate spatial patterns brings more possibilities for revealing capacity reserves in the structure. On the other hand it is also evident that in the complex spatial pattern the reaction of individual elements to loading will have a spatial character, with six components of the internal forces. This can mean that in some elements of the structure there is greater stress than was determined for this element using the simple calculating pattern.

The contribution of spatial calculating models is especially significant when the real geometry of the structure, including its imperfections and defects, position and extent of loading are all taken into consideration, as well as other influences that are manifested by the redistribution of internal forces in the structure. This is possible only with the help of complex spatial patterns while in some cases the application of second-order patterns is needed to specify the real activity of the structure. It is very important to consider effects of elastic redistribution of internal forces when modelling imperfections and defects of elements of a structure, for this always provides an additional factor not included by approaches that consider such effects only with respect to the corresponding resistance of the element. When it is necessary to chose a second-order calculating pattern, the problem of entering the shape and size of initial imperfections in the elements of the structure is simplified in the case of existing structures. These values can be measured directly on the structure, so the geometric shape of the structure can be modelled in accordance with reality. If there are no major defects in the structure, we can again expect a significant contribution from this calculation because in the design each element of the structure is assessed according to the values for imperfections in their most dangerous form. However, there is a little likelihood that these values will be reached in all elements of the structure.

In case of any doubts about the correctness of the chosen calculating pattern, it is possible to verify it by test loading the existing structure. This procedure is rather rare in ground constructions, but is more often used on existing bridge structures, where dynamic as well as static loading tests can be used.

4.8.5   Intentional use of the plastic properties of steels

Due to long-term monitoring of the quality of manufactured steels there is now enough data for the statistical processing needed to provide probabilistic guarantees of required quality increases. In cases where certain assumptions about physical and mechanical properties of steels can be guaranteed to be met, it is safe to use their plastic reserve. Accordingly, the current European standards removed restrictions on the use of the plasticity approach to the calculation of cross-sectional current resistance of steel elements, compared with the more frequently used elasticity approach. This latter approach predominated in the past, so the majority of existing steel structures were designed in accordance with the theory of elasticity. When using the plasticity approach to the determination of cross-sectional resistance of existing steel structures, another reserve of reliability can be used. The basic conditions for intentional use of the plastic reserve of steels are prescribed by standards for designing steel structures [38], [39], [40]. These deal with material, stability and constructional conditions. With respect to the mechanical properties of the material, the conditions for the ratio of tensional strength to yield value f_y is $f_u / f_y > 1.2$ and for elongation $A_s > 15\%$. Regularly used steels such as S 235, S275 and S 355 meet these conditions. With respect to stability criteria, the theory of plasticity can be used only for cross-sections in which the elements do not swell, i.e. cross-sections of the first and second classes. At the same time, with respect to global loss of shape stability, in the places where there are plastic hinges, it is necessary that the elements and structures are secured diagonally against lateral deflection. The condition that the members of variable cross-section have wall and compression flanges of the first class with minimum length equal to double the height of the cross-section from the location of the plastic hinge, is added to the constructional requirements. It is also required that the cross-sections at the location of assumed plastic hinges are loaded in the plane of symmetry and are reinforced cross-wise.

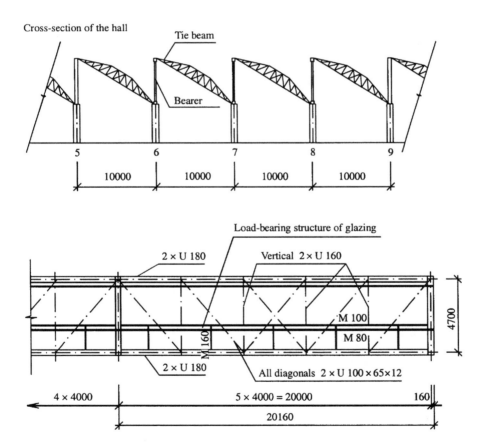

Figure 4.47 Schema for the roof structure and load-bearing bearer

If those conditions are met, it is possible to use the plasticity approach to determine the cross-sectional resistance of steel structures. In the case of existing structures it is necessary to check directly on the structure whether the given requirements have been met. At the end of this chapter we give one example of an actual application – a search for reserves in the structure according to the procedures described above. This example deals with assessment of the reliability of load-bearing roof bearers of an assembly hall. Lattice girders of 20 m spans are located on the columns. These are one-way system girders with double diagonals in the central lattice. They carry intermediate beams of 10 m span alternately bedded on the upper and lower chords of neighbouring bearers (see Figure 4.47) that are located on both chords away from joints so their reactions exert a local bending of the chords. The roof bearers have a parabolic upper chord and they constitute a one-way system with upward diagonals.

Glazing part of the vertical area of bearers, provided lighting for the working area of the hall. Because of damage to the hall by fire, the reliability of all elements of load-bearing structure of the roof was assessed, and insufficient capacity in the upper and lower chords of the bearers was discovered [48]. Because extensive refurbishment would be required to remedy the insufficient resistance of the bearers, the method of searching for reserves in the load-bearing roof structure [49]

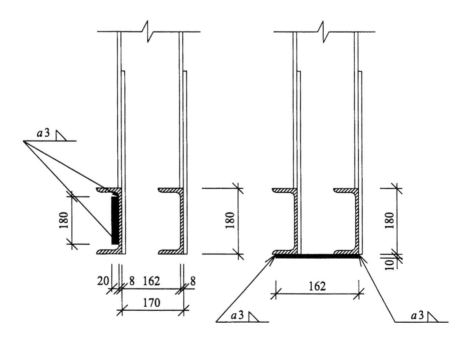

Figure 4.48 Alternative proposals for the reinforcement of the upper chord of a bearer

was chosen. Individual layers of the roof deck were measured, and from specimens taken the volume weight of individual materials was determined, which made
it possible to define more accurately the permanent loading exerted by the roof
deck. The weights of roof trusses and bearers, glazing, suspended ceiling and non-
glazed parts of the bearers were also determined in detail. The air conditioning
fittings were hung from the lower chords of bearers, so their weight was also determined by calculation. Variable short-term loads were considered in accordance
with corresponding standard [50]. The improved values for the loading were used
in the calculations for the bearer, which was modelled by a planar calculation
pattern with flexible strong chords and verticals. Diagonals joined by means of
hinges were considered. Also considered was the co-operation of the load-bearing
structure of the bearer's glazing as in Figure 4.51. To determine the resistance of
the suspect cross-sections of the lower and upper chords of the bearers, the plasticity approach was used because the conditions for its application had been met.
Subsequent assessment of reliability revealed the need to reinforce the lower chord
in the middle lattice of the bearer. Two alternative proposals for reinforcement of
the lower chord were prepared in accordance with the possibility of free access to
the lower chord. Both alternatives are shown in Figure 4.48.

Chapter 5

Repair and reconstruction of civil structures

5.1 Hall structures

5.1.1 Reconstruction of the roof structure of a stadium

During extensive modernization of the Winter Stadium in Bratislava, its steel roof structure was reconstructed. The original reinforced concrete skeleton with plan dimensions of 70.0 × 100.0 m was built over the period 1943 – 52. The raised seating around the ice rink was covered by a reinforced concrete decking with an 8.0 m overhang supported by a continuous girder in turn. This was supported by round reinforced concrete columns located 10 m apart. The load-bearing system for the raised seating was a transverse frame (Figure 5.1).

In 1957 the stadium ice rink was roofed [51] according to the proposal by Professor A. Tesar *et al.* which used a system of flat two-hinged steel-frame cross bonds of 52.8 m span (Figure 5.1). The horizontal forces on the two-hinged frames were absorbed by a closed horizontal lattice stiffener (Figure 5.2) suspended from an inclined draw bar on the frames. The columns under the new cross bonds were reinforced by ϕ 420/12 steel pipes. The space between the original reinforced ϕ 350 mm concrete column and the steel jacket was injected with cement mortar. Longitudinal strength was secured by stabilizing frames. The axial distance between the roof trusses was 5.0 m. The load-bearing underlayer of the roofing material, timber formwork, was bedded on trusses at an axial distance of 1.67 m. The roofing itself was made from double hardboard protected by ruberoid coating.

Before the planned reconstruction of the structure in 1987, specialists prepared an expert assessment of the steel and reinforced concrete load-bearing structure [52], the roof deck and the foundation structures. They found that:

- The structure of the roof deck and lighting in the plane of the inclined draw bars did not meet the spatial requirements for its micro-climate and that heat insulation should be provided to improve the energy economics of the building
- The original cross bonds – flat two-hinged steel frames with 52.28 m span – would not meet the requirements of the new loadings from the heat-insulated roof and maintenance footbridges

- The reinforced concrete supporting structure as well as the foundation structures were able to carry the increased loadings.

In the proposal of reconstruction it was necessary to take into consideration the following requirements:

- Replace the load-bearing underlayer and original roofing material
- Fit heat insulation to the roof deck
- Realign to verhaul the external window belts, originally located on an inclined plane
- Retain and use the original roof structure as far as possible
- Instal transverse maintenance footbridges at a distance of 10 m from the longitudinal axis of the structure.

Figure 5.1 Original steel roof structure after removal of the roofing material

Figure 5.2 Stiffener for fixing the response of the two-hinged frames

It was assumed that the original frame structure would be retained and used during reconstruction. According to the proposal by Stavoprojekt Bratislava, both transverse and longitudinal maintenance footbridges were to be suspended by inclined draw bars from the original frame structure below the level of the brackets of the reinforced concrete decking. However, the flat steel two-hinged frames proved not to have a sufficient reserve for the additional loading and so this solution would have reduced the height of the lighting of the hall.

The actual reconstruction was done according to Figure 5.3. The static system was changed with, the two-hinged frame being replaced by a lattice tie beam. The original frame (welded I cross-section) forms the upper chord and the horizontal stiffening members at both ends of the frame form part of the lower chord of new planar tie beams. The theoretical heights of the middle tie beams is 6.44 m and that of the outer tie beams in ties 1 and 6 is 5.54 m.

The internal forces of the cross bonds were determined in two stages. In the first stage, the cross bond acted as the original frame structure (the inclined suspensions were temporarily retained, so that the stiffeners taking over the horizontal reactions of the two-hinged frame could be suspended from them), which was loaded by the weight of the original structure after removal of the roofing material, the weight of its supporting base, and at the weight of the skylight structure. After these parts had been removed, a new lower chord, together with the skeleton of the transverse footbridge, were suspended by means of diagonals. Then the connected, so that the original two-hinged frame structure was thereby changed into additional lower chord with the draw bar of horizontal stiffeners was a lattice girder with a 52.28 m span.

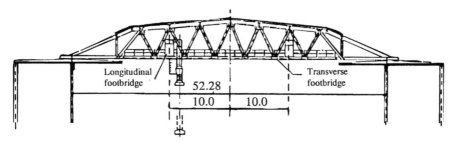

Longitudinal footbridge

52.28

10.0   10.0

Transverse footbridge

Figure 5.3 Schema for reconstruction

In the second stage, the cross bonds at distances of 10 m acted as lattice girders. They were loaded in consequence of removing the original inclined suspension of the main stiffener, and by the new roof deck and a glass-covered vertical wall, as well as by a part of the weight of the transverse footbridges, lights, exhaust installations and television cameras, with an occasional additional loading from maintenance footbridges, and from climatic effects. Due to the great rigidity of the main stiffener, firm and sliding supports were assumed in the calculation of internal forces. The rigidity of the support provided by main stiffener decreases in the direction from the outer bonds to the middle bond.

The additional lower chord was designed as a closed cross-section with a 280/16 wall, on which I 260 profiles were welded from both sides. Due to their slenderness the additional diagonals were made from tubes. The welded connection of the member of the original horizontal stiffener and additional lower chord of the binder is shown in Figure 5.4.

Figure 5.4 Connection of the additional lower chord with the original horizontal stiffener

The transverse maintenance footbridges are designed to be located at the level of the lower chord of tie beams. The width of footbridges is $1.03 + 0.71 = 1.74$ m and their skeletons consist of IPE 100 binders, or I 80 ledgers. The binders are welded to the lower chord of the tie beams. In the plan of the roof, the longitudinal footbridges are placed symmetrically at an axial distance of 20 m. The total length of the footbridges is 82.28 m. In the outer fields, footbridges with a span of 9.85 m were suspended under the original reinforced concrete decking brackets. The internal fields are suspended from the adopted cross bonds (Figure 5.5). The footbridges constitute a 1.98 m high lattice structure, with grid floors.

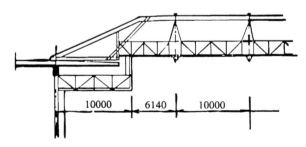

Figure 5.5 Schema for longitudinal footbridges

Figure 5.6 A view of a tie beam
and transverse footbridge

Figure 5.7 The roof structure after
refurbishment

The adapted cross bond – lattice tie beam is shown in Figure 5.6, and the roof structure after reconstruction in Figure 5.7.

In summary, in reconstructing the steel roof structure, an unusual but effective method was used, namely changing the static system of the structure. The total steel consumption was 106.3 tonnes, and of that 12.8 tonnes were used for grid floors.

5.1.2   *Reconstruction of a suspended roof*

Next is an example of the reconstruction of the suspended roof of the Stavoindustria experimental hall in Bratislava, Slovak Republic [52]. The experimental hall serves as the main warehouse (Figure 5.8) and is covered by a suspension roof with plan dimensions of 14.0×42.169 m.

Figure 5.8 The structure before reconstruction

During an assessment inspection it was found that there was excessive leakage in the roof. The leakage was caused by damaged hydroinsulation and insufficient maintenance. At the centre of the span of load-bearing cables (at the lowest points of the roof) there were high-strength patented forming wires subject to a permanently wet environment and in this area an uncontrolled corrosive process was operating. The original load-bearing cables of 30 mm diameter were made from $n \times \phi$ 4.5 – 5.0 mm high-strength patented forming wires. The axial distance between the cables was 1200 mm (Figure 5.9). Load-bearing cables carrying Siporex panels were embedded in concrete above a metal base of width approximately 150 mm.

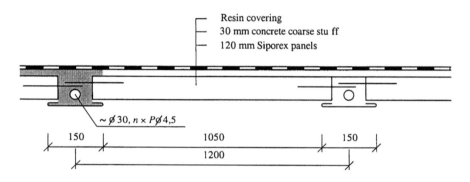

Figure 5.9 Composition of the original roof deck

The cables are anchored by concreting into the end reinforcing girders of a 195 mm thick continuous reinforced concrete plate. These 2100 mm wide reinforcing plates are supported at three points (at the ends and in the middle) in the straddled vertical column supports (Figure 5.10). The maximum height of the upper edges of the reinforcing girders is 11.725 m.

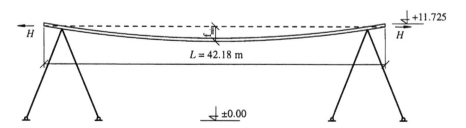

Figure 5.10 Schema for the suspended roof

Excessive corrosion of the patented wires was found by a probe made where a load-bearing cable existed from the reinforcing concrete plate. The end-supporting structures (reinforced concrete plate and straddled columns) were in good technical condition and could be retained for further use. The initial proposal for reconstruction of the roof assumed replacement of the suspended roof by a classical design – steel tie beams placed on the columns in the side walls of the structure. The proposal actually adopted preserved the original roof with minimum additional loading by using new trapezoid plates on additional cables (Figure 5.11).

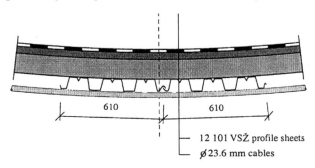

12 101 VSŽ profile sheets
$\varnothing$ 23.6 mm cables

Figure 5.11 Supporting of the original roof deck

The new suspension cables were located in the axes of the original cables are anchored in the end reinforcing reinforced concrete plates. The additional anchor device is shown in Figure 5.12.

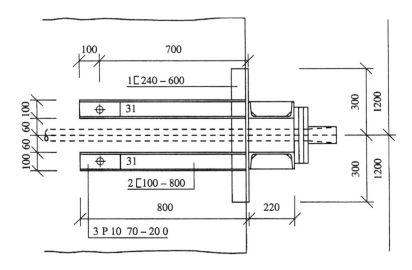

Figure 5.12 An anchor device for the new cables

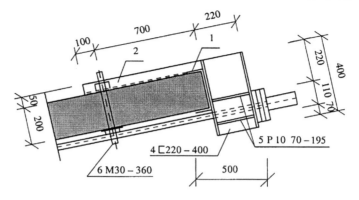

Figure 5.12 An anchor device for the new cables

The new load-bearing cables of the roof including supporting trapezoidal plates are shown in Figure 5.13.

Figure 5.13 A view of the new load-bearing cables

5.1.3 *Repair of the roof structure of a chemical plant*

The roof structures of several buildings in a chemical plant had suffered considerable damage caused mainly by a strongly corrosive environment acting on steel structures whose constructed design was unsuitable for such an aggressive environment because it did not allow for effective maintenance.

Tie beams and outer trusses were located on external load-bearing masonry. The corrosion losses on tabular surfaces reached 20 to 50 per cent, and there was also heavy corrosion in areas where steel structures rested on load-bearing walls.

Within the framework of the expert assessment of the technical condition of the steel load-bearing structures of the roofs, some proposals for repairs of the roofs were devised on the basis of the results of detailed assessing inspection.

The original steel structural systems of lattice girders with articulated members had been considerably weakened by corrosion, and the lower chords of tie beams were deformed locally by additional loading from distributional piping. In its present condition the load-bearing structures did not meet strength requirements and their service reliability and safety were not secure.

Figure 5.14 The original roof structure

Figure 5.15 The new roof deck

The roof structure of the hall (Figure 5.14), with plan dimensions of 34.5 × 39.2 m, comprised three identical lattice direct-chord tie beams, roof trusses, and stiffeners located between the tie beams. The tie beams and outer trusses were located onto the external masonry. While corrosion losses on tabular surfaces reached 20 per cent, the heaviest corrosion was in areas where steel rested on load-bearing walls. For this hall, the following repairs were carried out:

- The roof deck was replaced by a new one (Figure 5.15), which was lighter included a load-bearing roofing underlayer
- The lower chord of the outer trusses and the walls of trusses in areas where they rested on the walls were reinforced.

- The inclined struts in the outer fields were supplemented by a new strut.

The case of the hall building of plan dimensions 19.85 × 67.5 m was more complicated. The hall is covered by saddle roof structure. Lattice tie beams of triangular section with subsidiary compression verticals rest on load-bearing external walls. The axial distances in plan vary; places where there was the greatest corrosive damage (Figure 5.16), their axial distance was 5.8 m. Longitudinal stiffening was incomplete. In the course of long-term operation, substantial corrosive weakening of tie beams and trusses had occurred, and thereby the integrity and functioning of the aluminium suspended ceiling and the whole roof deck were impaired. The roof trusses at the drip were considerably damaged by corrosion with losses reaching as much as 50 per cent.

In this hall, the following repairs were carried out:

- The tie beams were reinforced: (the upper and lower chords had their loading from technical equipment removed, and new verticals were added
- A vertical stiffener was added between the tie beams.

Figure 5.16 Corrosion damage to girders bedded in load-bearing masonry

Figure 5.17 Replacement of the original roof deck

Figure 5.18 The new truss system

These repairs were done while the original roof deck was still in place. After their completion the roof deck (Figure 5.17) and trusses (Figure 5.18) were replaced, and the structure was completed by fitting roof stiffeners.

The repairs to the steel roof structures were carried out in confined conditions, without interruption to operations in the adjoining halls. The situation with the other halls was dealt with later in a similar way, i.e. by reinforcing the tie beams as well as the steel columns (Figures 5.19 and 5.20), and eventually by replacing entire tie beams with new ones.

Figure 5.19 Corrosion damage to columns

Figure 5.20 Reinforcement of the original columns damaged by corrosion

5.1.4   Reinforcement of tie beams in a cinema theatre

This case involves reinforcement of direct chord lattice tie beams of tubular structure with span $L = 12.0$ m and height $h = 1.20$ m, to which, large deformations of
web members had occurred after erection and partial loading (Figure 5.21).

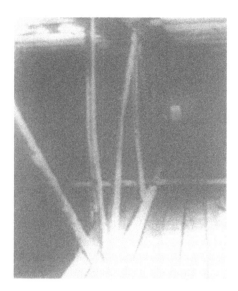

Figure 5.21   Deformation of web members of the tie beams

From the overall evaluation it was established that the unserviceable condition
of the tie beams was the combined result of many unfavourable factors related to
design manufacture, and erection procedures. Deficiencies of the project included
insufficient dimensioning of the compression diagonal D_4, and unsuitable cross-
section (tubular) of the upper chord, which was also dimensioned insufficiently,
both deficiencies evident in the manufacturing drawing.

The manufacturer changed the designed dimensions of the upper chord (ϕ 89.4)
without the designer's approval to ϕ 83, while the wall thickness had been changed
from 3.3 to 4.15 mm (as determined by ultrasound apparatus). The first and last
nodes of the lower chord were incorrectly made with one-sided plate (Figure 5.22),
and the quality of connecting welds was not good. When the tie beams were received, the proper checking procedure were not followed. The finished tie beams
were not compared with the manufacturing drawing, and the differences were not
recorded in the documentation; in good practice each change must be checked by
calculation and its appropriateness must be demonstrated. Incorrectly made tie
beams should not have been used in construction. It was also established that large
initial deformation of some web members resulted from faulty handling in storage

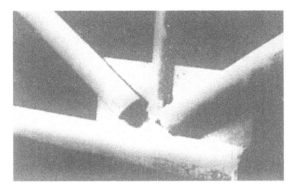

Figure 5.22 Edge nodes of the upper chord of the tie beams

and in the erection of the tie beams. The deformation of compression members was found to increase with graduated loading.

On the basis of expert opinion, reinforcement of the tie beams was proposed according to the scheme in Figure 5.23. New nodes were created according to details A and B. The reconstruction was carried out with full unloading of individual tie beams, achieved by lifting off the covering. In this unloaded condition, the members were aligned, the tie beam was completed by new members, and edge nodes of the lower chord were modified.

By adding new members, the buckling lengths of diagonals D_4 and the upper chord were reduced, so that the stresses on them fell to admissible values.

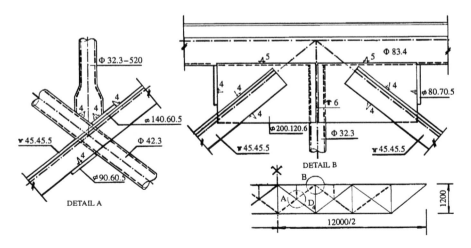

Figure 5.23 A schema for the reinforcement of the tie beams; new nodes were created according to details A and B

5.1.5   Reinforcement of the cross beams of horizontal frames

This case involved the reinforcement of the cross beams in the horizontal frames of a customs post. The roofing used a Maculan system in which the couplings of high-strength prestressed wires are anchored in horizontal two-hinged frames in the roof plane. The reactive forces in these frames are distributed via longitudinal struts. Because it is a welded structure with difficult erection welds, the project required radioscopic checking of welds located in exposed places. The examination of the welds by the Research Institute of Welding in Bratislava showed that these were of insufficient quality. In the prevailing conditions, prestressing was carried out while the actual stress conditions in the exposed cross-sections of the steel structure was monitored.

The measurement taken led to conclusion that where the posts of horizontal frame were connected, local plasticizing had occurred. This resulted in a rearrangement of the bending moments in the central part of the cross-beam and increased sagging of the cross beam. The cause of this behaviour of the structural frame was identified as the reduced capacity of the low-quality connecting welds.

In order not to increase further the stresses in the most-stressed parts of the structure, which would lead to further deformation of the cross beams of the horizontal frames and hence by an undesirable reduction in the forces in the prestressed high-strength wires (which in turn would interfere with the correct functioning of the roof area), it was decided to support the stressed cross-beam frames by means of struts, as shown in Figure 5.24. The reinforcing struts were connected to the cross beam according to detail; this would prevent from heat from the tensioned table effecting the cross beam. The base plate is connected only by weak welds. A general view of the reinforced frame is given by Figure 5.25.

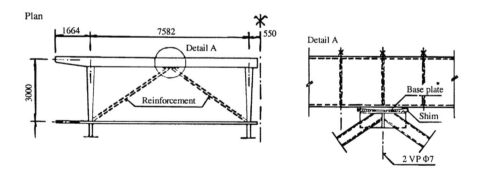

Figure 5.24 Schema for supporting the cross beam of a horizontal frame, together with details of the connection of the strut to the cross beam

Figure 5.25 A general view of reinforced horizontal frame

5.1.6 Reinforcement of roof deck of a hall

The load-bearing structures and their technical condition
before reconstruction

The roof over a garage and assembly hall was designed as a single deck with internal drainage via roof inlets. The roof deck, also houses the lighting space through the use of suspended ceilings. The overall dimensions of the roof are 145 × 72 m, with some dilation towards the middle of the larger dimension.

The load-bearing structure of the hall was designed as a system of cross bonds formed by pre-cast reinforced-concrete columns of section 60 × 60 cm and steel lattice tie beams. The span of the tie beams is 18 m. The distance between the columns around the periphery is 6 m and the distance between the internal columns of the cross bond is 12 m. The columns of the cross bond are connected by bearers on which rest intermediate lattice tie beams of 18 m span. The steel structure was in good technical condition.

The load-bearing part of the roof deck is made of porous concrete armoured roof UNIPOL panels with a span of 6 m. These are of gas-silicate type, and were manufactured in the early 1970s to widths of 0.8 m, 1.2 m and 1.6 m, and constant height of 240 mm. These panels rest on the upper chord of the tie beams, which have a bearing width of 81 mm. Where ventilators or other fittings pass through the plane, reinforced concrete plates are used instead of the porous concrete panels. On the other side, porous concrete pre-cast girders of 6 m span rest on the intermediate steel lattice tie beams.

After detailed analysis of the technical condition of the porous concrete roof panels it was concluded that their load-bearing capacity was exhausted and that in some parts of the hall their state constituted an emergency situation. The most damaged parts of the roof deck were immediately given temporary support (Figure 5.26).

Figure 5.26   Temporary support of the roof panels

In the reconstruction proposal, it was assumed that sound porous concrete roof panels could still be used, and their capacity would be increased by changing the static load-bearing system. The roof structure was supplemented by new reinforcing elements, the porous concrete panels were secured to prevent them falling out, and the strength of the roof plane was increased.

Reconstruction of the steel load-bearing roof structures
The customer needed to ensure the operational reliability and prolong the life of the roof structures with minimum intervention to the original roof deck, despite the fact that in some areas (Figure 5.27) panels fell out during operation, notably at inlets where there was water overloading.

Figure 5.27   Modification in areas where roof panels had fallen

From static calculations, it was found out that the upper chords of tie beams and longitudinal reinforcement (Figure 5.28) were insufficiently dimensioned given the overall conditions obtaining in the roof deck. In addition, on some tie beams the suspended one-tonne crane had caught in the joints of the upper chord. Given these findings, it was necessary to design a support system that would shorten and secure the buckling length of the upper chords of the tie beams vertically from their plane to a height of 3 m.

Figure 5.28 Original longitudinal reinforcement of the girders under the roof deck

Figure 5.29 The new system of longitudinal and transverse girders under the roof deck

In the plane of the upper chords of tie beams, a new load-bearing system of longitudinal and transverse girders (Figure 5.29) was proposed, which would redistribute short-term accidental loads from the roof deck. It also forms a supporting system that would prevent porous concrete panels from falling out by providing an underlayer and filler.

The original vertical reinforcement between the roof trusses was dismantled and replaced by strut reinforcers. The reinforcing system was completed by transverse and longitudinal chords of reinforcement (Figure 5.30) located under the plane of the roof deck. Some of the material from the dismantled vertical reinforcement could be used here.

Figure 5.30   The new window system under the roof deck

All of the new structure was designed bearing in mind the possibility that the roof deck might later be replaced by a new, lighter one.

The reconstruction was completed within five months under full production operations in the hall. In the original steel structure, 47 tonnes of reinforcement was dismantled, and of that material 35 tonnes were used in the newly designed reinforcing system. In the roof structure, 130 tonnes of new rolled steel and gusset plates were fitted.

5.1.7   Reconstruction of a metal foundry

This metal foundry was built at the end of the nineteenth century. The original roof structure was wooden, bedded on external brick masonry. Nailed tie beams with span of 25 m were spaces 5 m apart, and a skylight of width 5 m and height 2.2 m was located in the longitudinal axis of roofing. On the basis of a detailed examination of the roof structure and static calculations for the wooden tie beams, and the bracing in the roof structure, it was found that part of the original structure was in emergency condition due to overloading by fly-ash to depths ranging from 100 to 150 mm that had not been removed regularly, and by increasing number of layers of cardboard roofing used during maintenance of the roof deck. It was also found that the original roof structure did not have bracing against cross-winds, and many members of longitudinal bracing in three vertical planes located on the longitudinal axis of the roofing and the external vertical of the skylight were damaged or had been removed in course of operations. The total stability, as well as the local stability of the main load-bearing beams of the roofing was thereby seriously reduced, and because of the critical loading of the roof by the layer of fly-ash, all compression chords of the tie beams were buckled out of their true plane.

Restoring the reliability of the complete roof structure required a number of modifications and are marked in Figure 5.31 by the darker lines:

- Secure the stability of the roof structure by fitting cross-wind bracing in three areas of the original wooden part of the roof and in one area of the new steel part.
- Reduce the stresses in the chord members of the original wooden tie beams by supporting them with steel posts (Figure 5.31 section cut B – B) Such support would reduce the stress of the chord members below the calculated values, and at the same time would reverse the direction of the axial forces in the area of the support compared to the diagonal direction in the diagonal members of the tie beams. The reinforcement of these diagonals was achieved by using nailed wooden stiffening plates.

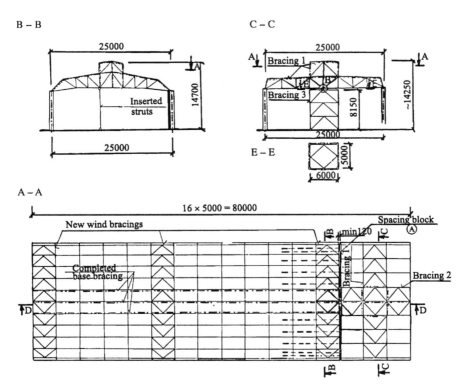

Figure 5.31 Arrangement of the roof structure of the foundry after reconstruction

- Repair or replace vertical bracing in three planes that transfer longitudinal forces to the vertical tower bracing (Figures 5.31 sections C – C and E – E) located so that production processes in the hall of the foundry would not be interrupted.

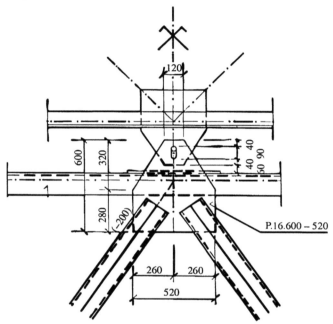

Figure 5.32 A detail of the connection of the roof structure to the vertical bracing

Connecting the bracing through the vertical opening means they carry only horizontal forces, so the role of the tie beam is not compromised by vertical loadings.

5.1.8   Reinforcement of the steel structure of a shipyard hall

Before modification, two electric bridge cranes with lifting power of 25 and 5 tonnes (cranes No.22 and 23) were operating in hall No.15, with the possibility of coupling the cranes so they operating together. Due to the increasing size and weight of assemblage parts of ships, replacement of the 25 tonne crane in hall No.15 with a new bridge crane with increased lifting power of 50 tonnes was considered. After some modifications to the hall, two bridge cranes with 25 and 50 tonne lifting power were put in operation (Figure 5.33).

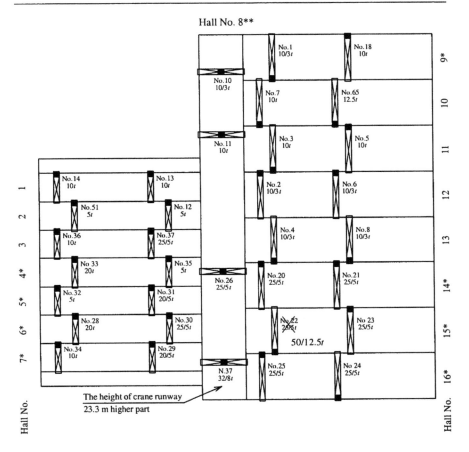

Figure 5.33 Arrangement of bridge cranes in the halls

Neither the construction documentation for the original project and the modifications nor the static calculations for the load-bearing structure of the hall were available as background for the preparation of the expert's opinion. In order to assess the technical condition of steel load-bearing structure, and to provide the necessary background, the builder's workers carried out detailed assessing inspections in 1998. They measured the basic dimensions of the structure and determined the cross-sections of load-bearing elements, mechanical damage and corrosion losses, particularly those to the columns and bracing of the steel structure. A schema for the cross bond in hall No.15 is given in Figure 5.34. The axial distance between cross bonds is 15.0 m.

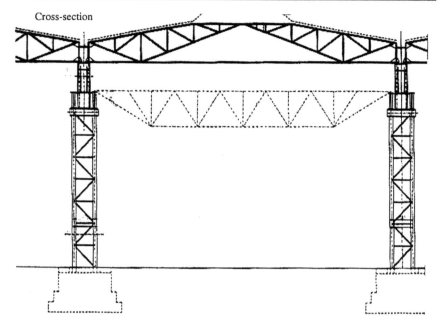

Figure 5.34 A schema for the cross bond in hall No.15

The technical condition of some columns, cross bonds and bracing members was unsatisfactory due to excessive corrosive damage. Near passages through the floor, parts of the cross-sections of the columns were totally corroded (Figure 5.35). Similarly, the members (chords) of the bracing stiffeners (Figure 5.36) were totally corroded, and could not be used any more.

Figure 5.35 Corrosive damage to a column

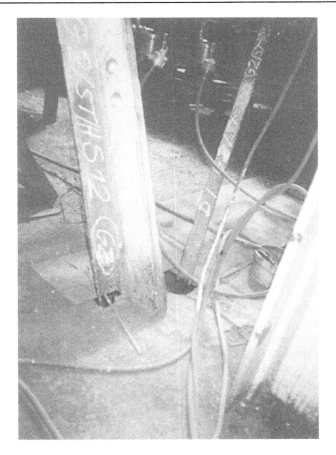

Figure 5.36 Corroded members of a bracing stiffener

However, from the assessing inspections, the technical condition of the crane run-ways was evaluated as good, without major mechanical or corrosive damage. The results of checking static calculations confirmed that the original cross-sections of girders of the crane runways met the strength and durability requirements for the increased loading from new bridge cranes with lifting powers of 25 and 50 tonnes, provided they were not coupled. The original wooden floor on the horizontal rein-forcing girders was replaced by a metal grid floor. The bodies of columns weak-ened by corrosion (Figure 5.37) were reinforced to restore their original capacity. After cutting away the concrete from outside the existing tables, additional P 25 reinforcing tables were welded to them by means of fillet welds. The minimum length of the tables is $2 \times 300 = 600$ mm.

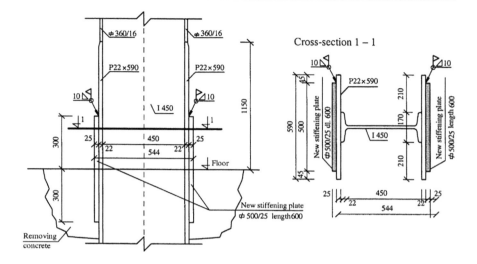

Figure 5.37 Local reinforcement of a column where it is connected to the floor

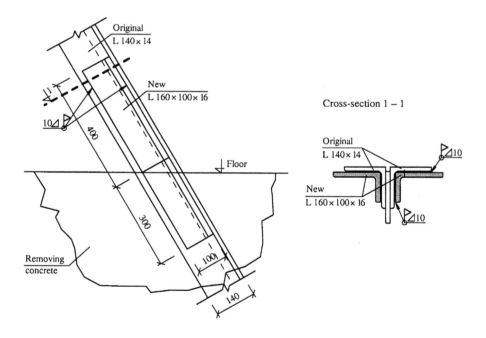

Figure 5.38 Reinforcement of a member of a bracing stiffener

Members of bracing stiffeners that had been damaged by corrosion (particularly their chords) where they were connected to the floor were locally reinforced. After removing concrete from the surroundings of the damaged member, additional reinforcing angles or split stiffening plates were welded onto the original angles. The minimum length of the reinforcing elements was 60 mm (Figure 5.38).

All column bases were covered with concrete shaped to shed water in order to protect the area where the columns and stiffeners were connected to the floor. The minimum height of the concrete covering on the column bodies was 200 mm and on the stiffeners 300 mm (Figure 5.39). Due to the very poor condition of the concrete, new anticorrosive coatings were applied to the entire steel structure of the hall after reinforcement was completed.

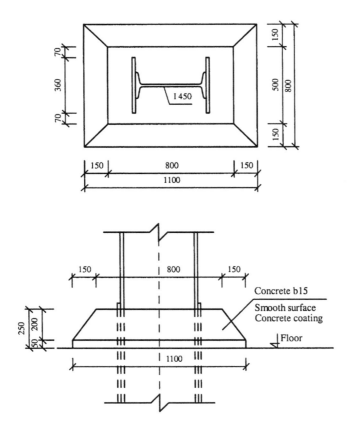

Figure 5.39 Concrete cover of column bodies

5.1.9 Refurbishment of the steel structure of a coal-cleaning plant

When assessing whether older buildings with steel load-bearing structures can safely carry their loadings associated with new technological processes, it is usually found that the assumptions of the original calculations that divided the structure into individual planar load-bearing systems which carry partial loads over the total loading width, do not allow for significant increases in the original loading of the structures. Therefore in proposals for refurbishment such strucutures it is necessary to deduce from the calculations a model that makes it possible to describe as accurately as possible the internal forces involved in the operation of the structure.

During the examination of the present condition of the steel structure of the building, which dates from 1937, the load-bearing system was compared to the relevant drawings with respect to the columns, bearers, platforms, stiffeners, etc. Close attention was paid to the location and completeness of the stiffeners, which were essential to the spatial calculation model for the structure as a whole. It was found that some elements or parts of elements of the steel structure had either not been fitted or they were removed later.

The loading of the steel structure compared to the original values assumed by the project was found to be much higher. In many places this was caused by storing spare or waste material, or by accumulation of raw material. However, because in the case of a building like this part of the total loading significantly affects the fatigue strength of constructional parts and joints (via not only pulsating stresses, but also in some cases alternating stresses). Therefore special attention was paid to dynamic loads that caused vibration and noise.

The entire structure operated in a corrosively aggressive environment. The lower parts of columns and their bases in many places were permanently in the mud. Eventually these areas became polluted and clogged with material. The total loss of cross-sectional area in parts of steel structure due to corrosion was estimated at 10 per cent, and up to 15 per cent in the more damaged parts.

The calculations for steel load-bearing structure in the original project assumed that vertical loadings were carried by cross beams and struts of the load-bearing skeleton, and horizontal loadings by lattice stiffeners (Figure 5.40).

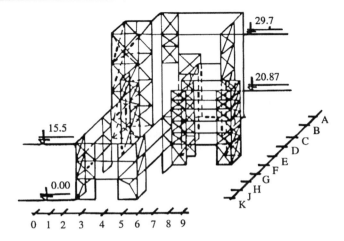

Figure 5.40 Schema for the arrangement of vertical lattice stiffeners of the steel load-bearing structure

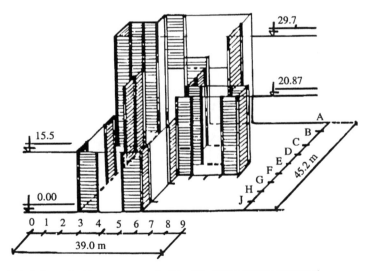

Figure 5.41 Schema for the arrangement of the thin-walled spatial system of the steel structure

As the reinforced-concrete working platforms at individual floor level are represented by rigid horizontal stiffeners, for the horizontal loading a thin-walled spatial system was chosen as the calculation model for the steel structure. In this system, the chords of vertical lattice stiffeners which for areas of stress concentration connected by diagonal members, were replaced in their function by full walls of equivalent thickness (Figure 5.41).

A uniform distribution of normal stresses over the column cross-sections and shear stresses over the cross-section of alternate solid walls was assumed. This means that in calculating the horizontal loading of the structure, only the columns that were parts of the stiffeners were considered.

The cross-section of the calculation model varies with the building height, so the position of the centre of torsion varies too. Therefore the line connecting the torsion centres of individual cross-sections is not straight but irregular. In the calculation model of the structure it was assumed for simplicity that the line connecting the torsion centres (the axis of the thin-walled system) is a straight line. In the calculations for the steel structure designed a thin-walled spatial system, first-order theory is adequate because the height of the building is small in comparison to the dimenssions of the cross-section.

The calculations of cross-section characteristics were done on the condition that non-deformability of individual cross-sections (at the level of floors) was secured by strong concrete columns according to the theory developed by V.Z. Vlasov. The rigidity of the columns in the structure was determined from the vertical loading (their overweight, any temporary weights), the horizontal wind load, and dynamical loadings caused by cranked and self-balancing sieves and transport conveyers. As well as the bending oscillation caused by such dynamic loading, the effects of torsion oscillation caused by eccentric oscillation was also examined with regard to the torsion centres of individual cross-sections in the calculations model for the structure. The calculations showed that the cause of the undesirable oscillation of the structure, in the region between column rows G and K (Figure 5.42) was insufficeint strength of the stiffeners in row K (Figure 5.43).

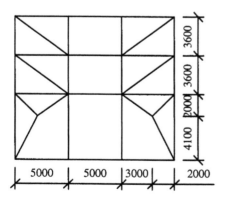

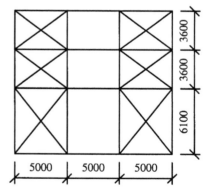

Figure 5.42 Arrangement of the original vertical stiffeners in column row K

Figure 5.43 Modification of stiffeners in column row K

From detailed assessment of all columns in the steel structure it was determined that the following columns did not meet the buckling or fatigue requirements:
- In the section from 0.0 to + 5.4 m: A4; B2; E2 (Figure 5.30)
- In the section from +5.4 + 9.3 m: B2
- In the section from +9.3 + 12.8m: B0; B2; B3; B4; B5; C0
- In the section from +12.8 to 15.5 m: B2; B3; B4; B5.

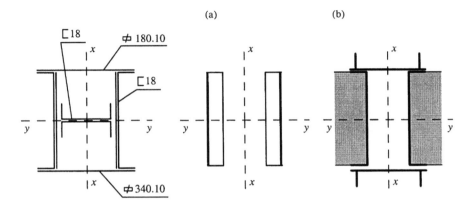

(a)	(b)

Figure 5.44 Original cross-section of column B2 in section 0.0 − +5.4 m

Figure 5.45 Ways of reinforcing the columns of the steel structure

There was proposed necessary reinforcement of cross-sections of the relevant columns by welding steel plates to the I-profiles of the original cross-section (in Figure 5.45(a)), or by welding such plates according to Figure 5.45(b) in those places where the filling masonry of the external wall did not allow the method shown in Figure 5.45(a).

By modifying the stiffeners through inserting new diagonal members and reinforcing the cross-sections columns using belts of vertical stiffeners, the dynamics acceleration of the load-bearing structure fell to a value of $a = 0.145$ m.s^{-2}, which is significantly lower than the damaging values of $a = 1$ m.s^{-2}.

The detailed analytic calculations of the case of a spatial steel load-bearing structure of this coal-cleaning plant showed that when conceiving proposals for load-bearing structures of industrial buildings with difficult dynamic service regimes, the arrangement of the individual stiffeners in the load-bearing structure must be optimised not only for static loading effects, but for dynamical effects too.

5.1.10 Additional buildings

A new triple-bay hall with a steel load-bearing structure was added to the original hall. The building recessed into the ground to a depth of 1.62 m. The footings for steel columns are designed as monolithic column bases 1.3 m high and made footings of B20 concrete. Under the external walls there are strip footings of B20 concrete 400 mm wide and 800 mm deep. A monolithic slab 150 mm thick of wire-reinforced concrete for a 30 kN/m² loading is also a part of the foundations. The load-bearing roof system consists of steel plates connected to steel trusses (Figures 5.46 and 5.47).

Figure 5.46 The original hall with a concrete load-bearing structure

Figure 5.47 The additional building with a steel load-bearing structure

The trusses were designed using rolled steel I 160 profiles, while the beams used tubular profiles of 12.255 m (outer bays) and 12.0 m (middle bay). The internal columns (Figure 5.49) are made from steel profiles HE 200 B. Anchorage is provided by 25 mm bearing plates and by M36.3 bolts pre-concreted into the base foundations. Hardening of the concrete in the hall was achieved by aeration via the tubular profiles.

Figure 5.48 A view of the tubular tie beams and skylight

Figure 5.49 The load-bearing steel columns

An additional building for the Fima Mobel department store, Dunajská Streda

The outer columns were located on the original base foundations to which the new base foundation base were linked at a depth of 2.2 m. This linking to the original foundation structure was realised by means of bearers of HE 260 B profile in 4.8 m modules and ceiling beams of IPE 220 profile of 1.2 m each (Figure 5.50). The brackets were of HE 260 B profile in the direction of the bearers and I 260 profile in the direction of the ceiling beams. On the steel ceiling elements (in 1 NP and 2 NP) stops were welded, each of L profile 900 mm. Also the steel ceiling elements were fitted VSZ 10 001 sheet of height 30 mm with a 60 mm concrete panel (i.e. the total thickness of the plate is 90 mm). The roof structure of the new building used IPE 300 and IPE 200 rolled steel profiles on HE 200 B steel columns. Wooden trusses of dimensions 80/200 mm and 50/200 mm used SI material and full formwork 24 mm thick.

Figure 5.50 Steel load-bearing elements of the ceiling of the additional building

Figure 5.51 Steel load-bearing structure of the additional building

The columns (Figure 5.51) are of steel HE 200 b profile. Anchorage of the outer columns in the original building was secured by anchor plates 40 mm thick and anchors HILTI, HVA M16 fixed to the foundation beds. Anchorage of the other columns was designed as a solid fixation with the columns together with anchor plates and stiffeners embedded within the frames of the reinforced concrete bases. The columns of HE 200 B profile of S 235 material were anchored in the existing reinforced concrete structures.

Reinforcement of the new steel building was achieved by aeration from the tubular profiles. In the vertical plane, reinforcement is also achieved by fixing the front columns to the existing reinforced concrete structure. In the horizontal plane, reinforcement was done using monolithic reinforced concrete plates and full formwork 24 mm thick. General views of the finished construction are given in Figures 5.52 and 5.53.

Figure 5.52 A general view of the finished building

Figure 5.53 Another view of the finished building

Roofing the atrium of the Drevomont production hall, Dunajská Streda

This case involved roofing a pre-existing atrium (Figure 5.54).The existing reinforced concrete columns had reinforced concrete foundation bases with plan dimensions of 1.5 × 1.5 m at a depth of approximately 2.0 m below ground level. Due
to the fact that in the original construction the outer bases used were of identical
dimensions to those of the more heavily loaded internal bases, the foundations
after the additional loading still met the requirements of the first and second limiting conditions, so it was not necessary to reinforce them.

Figure 5.54 The original building before roofing the atrium

Figure 5.55 Roofing of the atrium using tubular steel tie beams

The structure of the roof is designed using steel lattice tie beams 2.0 m high with 19.9 m span, anchored on steel columns (Figure 5.55). Steel web trusses of I 160 profile are fixed to the lattice tie beams at a distance of 2.0 m by gusset plates. Plate of RANNILA RAN 35 – 0.75 mm is fitted to the trusses. The vertical load-bearing structure uses the original reinforced concrete columns in the existing skeleton of the building, and reinforcement of the atrium structure acts in the plane of roof, secured by diagonal roof stiffeners.

The roof structure was built in 2001. Such structures present a number of unusual problems. All modifications, repairs, realignment, reinforcement and refurbishment of the steel structures must be done in ways that will secure the stability and safety of the structure.

5. 2 Refurbishment of steel structures after fire damage

Fire resistance of steel structures must meet the valid standards. Structures can be protected against fire hazard by proper constructional design and surface protection. In high temperatures fire (see Chapter 2 Section 2.4.3) the limit states in capacity and serviceability may be exceeded. What follows is an example of a light hall structure without fire hazard protection, whose capacity limits were suddenly exhausted by a fire. It collapsed and was irreparably damaged (Figure 5.56).

Figure 5.56 The steel structure of a light hall damaged by fire

In this case it was found that the effect of the temperature of the fire on structure depended mostly on the amount, properties and arrangement of the combustible material, and on the oxygen access and spatial arrangement of the building. The hall was originally built to store canned fruit. Because of poor crops, at the time of fire other combustible material from a factory manufacturing safety matches [51],

[52] was stored in the hall. The hall was a two-storey building with plan dimensions of 24.0 × 30.0 m, and the fire which began on the first floor reached the intermediate cross bonds V1 – V4. These cross bonds consisted of bipolar web welded frames with spans of about 2.0 + 12.0 m (Figure 5.57).

Figure 5.57 A view of the load-bearing structure for the first floor

Figure 5.58   Frame cross beam V1 in the south bay

After two detailed examinations of the extent of damage to the load-bearing structure was assessed.

The roof trusses had already been removed at the time of the examinations. The main load-bearing structure (frames and stiffeners) had suffered most damage in the south bay. The frame cross-beams were strongly distorted with the main components of distortion being horizontal bending and twisting (Figure 5.58).

This suggested that the load-bearing structure had not been heated to very high temperatures (over 450°C), so the deformations must have been caused by tensions resulting from uneven heating combined with the strength of the structure that prevented expansion. This was indicated by large deformations in the longitudinal stiffeners (Figure 5.59).

(a)   (b)

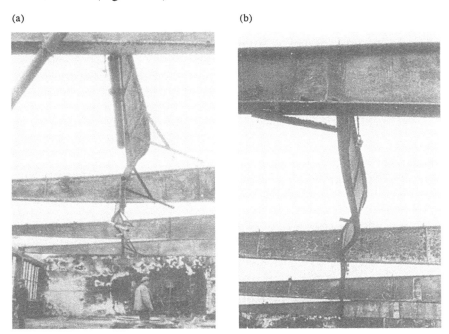

Figure 5.59 Deformation of the longitudinal stiffeners (a) in the south bay (b) in the north bay

In the south bay there were also large deformations in the lateral longitudinal stiffeners at the top of the window frames (Figure 5.60). The nature of these deformations – bowing of non-fixed beams again indicates that the cause was prevented expansion with raised temperature. The structural deformations were smaller in the north bay, but the longitudinal stiffeners were still significantly distorted.

The middle columns were displaced from the vertical position and slightly twisted due to the deformation of the longitudinal stiffeners. In the examinations, the deformations of columns were measured. The outer columns remained relatively straight however, and only at the column of frame V2 in the north bay was there bowing of two I-profile beams.

There was also a local fire on the first floor. The middle column V1 and the middle longitudinal stiffener between bonds 0 and 1 were damaged. The middle column was slightly bent (the maximum deviation over the full height from the floor was 20.0 mm) and in its lower region there was buckling of the column wall. The nature of this buckling and size of the deviation are shown in Figure 5.61.

In order to determine the effect of the fire on the properties of the steel work, a stretching test and a notch toughness test were carried out. Specimen material was taken from the lower chord of the middle longitudinal stiffener between bonds V2 and V3 on the first floor and from the lower table of the middle stiffener 0 – 1 on the ground floor. In the streching test, the yield value, ultimate strength, elongation, and contraction were measured.

The results of the test were as follows:

- The structure was made of 37(S235) steel
- The yield values and ultimate strength showed a wide spread but were within normal limits. Only the ultimate strength of specimen No.1 was lower than that assumed for S235 steel
- The elongation and contraction measures were relatively high for all specimens, providing evidence of that the material properties of the steel used were sound.
- The notch toughness of test bodies 1, 2a and 2b was relatively low, reaching only 40 to 50 per cent of the notch toughness of test bodies 3a, b and c.

Figure 5.60 Deformation of the beams at the top of the window frames between bonds V1 and V2 in the south bay

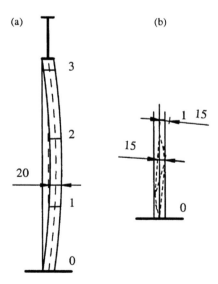

Figure 5.61 (a) Deformation of middle column V1 on the ground floor, (b) local deformation of the column wall

These results show that this is a case of steel effervescence, by which the heating at first floor to level to temperatures as high as 300°C caused texturizing to occur. At ground-floor level the heating did not reach such temperatures, so the decrease in notch toughness did not occur. The test results also showed that the steel was not affected by the fire to such an extent that its properties were changed. Therefore the parts of the steel structure that were not greatly distorted could be retained and could be welded satisfactorily during reconstruction.

Due to excessive overstraining, the upper frame cross beams and longitudinal stiffeners had to be replaced. The new upper frame cross beams P1 were designed as welded web I cross-sections of variable strength (Figure 5.62).

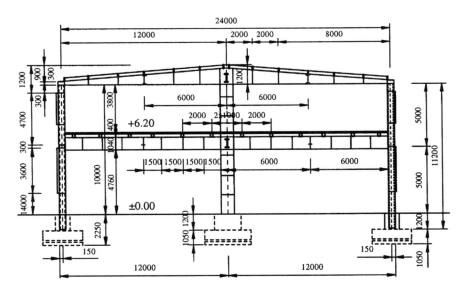

Figure 5.62   Schema for the cross beams V1 – V4 after reconstruction

Static calculations established that there was a sufficient reserve in the capacity of the columns capacity so the original columns were retained in the upper areas. The vertically distorted columns were realigned before the new roof structure was erected. To carry the required useful permanent loading of 4.0 kNm^{-2}, the floor girders of the intermediate ceiling had to be reinforced, as in Figure 5.63. To provide the required spatial rigidity during reconstruction, additional horizontal lattice transversal stiffeners with tubular web members of 60.5 mm diameter were designed. The one-storey production hall with large plan dimensions is covered by a sawtooth roof. Vertical steel lattice bearers of nominal height 4.7 m and span of 20.0 m were placed on the reinforced concrete columns. Inclined lattice girders with poly-

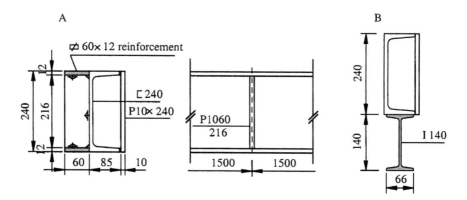

Figure 5.63 Proposed reinforcement of the floor girders

gonal upper chords and the span of 10.0 m were placed on the bearer chords. The suspended ceiling, which created by individual plates made of load-bearing grid of concreting steel mesh and 20 mm cement plaster, protected the steel load-bearing structure relatively well during fire. The three most damaged bearers and related roof trusses above the centre of the fire were replaced.

In the proposal for the reconstruction of the bearers it was assumed that both upper and lower chords of the retained bearers would be completely unloaded, so that these chords could be reinforced and the connecting rivets of the outer diagonal would be replaced by high-strength bolts. Due to lack of time, these bearer chords were not dismantled and so were not reinforced either. During reconstruction the bearer was supported at the node (Figure 5.64) by a column and tubes of 19.10 mm diameter.

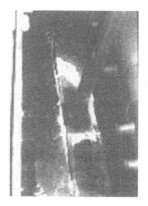

Figure 5.64 Support for the retained bearer

The internal forces acting on the reconstructed bearer were determined by taking into account its strength, geometric shape and static schema, as in Figure 5.65. The contact distance 1.0 m measured from the node was welded because of a shortage of high-strength bolts. All members of the bearer reconstructed in this way met the relevant strength requirements (Figure 5.65).

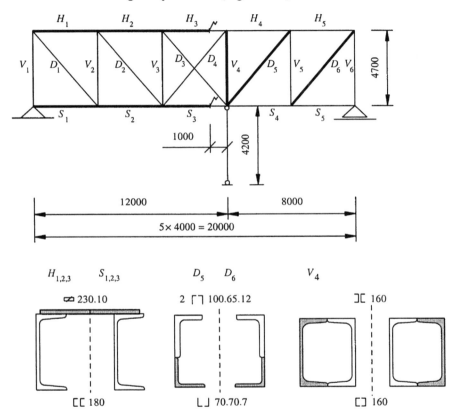

Figure 5.65 Reinforcement of the supported bearer

The case confirmed the great importance of protecting steel structures against high temperatures. Where structure was protected by a continuous suspended ceiling, relatively slightly damaged, even by a very strong fire. The material tests showed that the mechanical and technical properties of 37(S235) steel had not been substantially changed by the fire, even by fast cooling through quenching, so that in reconstruction it was possible to use welding. Careful assessment of the slightly deformed parts proved they had sufficient capacity and so could be retained in the structure. The extent of reconstruction was thereby significantly reduced and great time savings were achieved.

Chapter 6

Repair and reconstruction of engineering structures

6.1 Bridges and special structures

6.1.1 Cable bridge systems

6.1.1.1 Introduction

The number of cable systems has increased considerably over the last three decades thanks to their well-known advantages. Beside their economy, this has also resulted from the development of new architectural forms in the design of bridges and land constructions with large spans.

Along with their advantages, cable systems also have their disadvantages. Notable among these are their greater deformability, the danger of vibration, and thereby also the possibility of cable breakage, especially at and around the anchorage. Protecting cables against corrosion is more complicated than in the case of contemporary steel structures. In many types of cable system, when just a single cable breaks, this can lead to an unfortunate rearrangement of the internal forces that can cause collapse of the whole system.

6.1.1.2 Correcting the off-centering of the bridge of the Cabin Cable Railway

In 1988 – 9 the technical condition of the cable bridge of Cabin Cable Railway at Start in the High Tatras was assessed (Figure 6.1). At this bridge, incorrect feeding of the cables into the pulleys was a regular occurrence [64]. The expert assessment revealed that a steel structure had been wrongly erected, resulting in a linear slewing of the main girders of the bridge ranging from 8.0 to 169 mm in the plane of the plan (Figure 6.2).

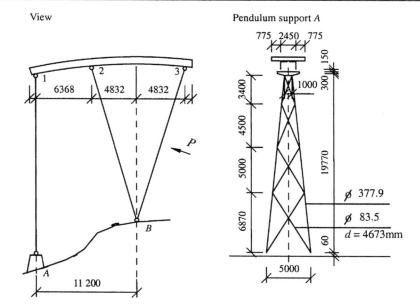

Figure 6.1   Schema for the cable bridge

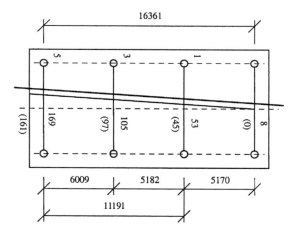

Figure 6.2   Slewing of the main girders

To correct the alignment of the saddle for the load-bearing cables and the pulley batteries of the tow cable, replacement and lengthening of the original supporting brackets for the suspension system was proposed (Figure 6.3). The cross-section, as well as the connection of original brackets on the lengthened side were reinforced.

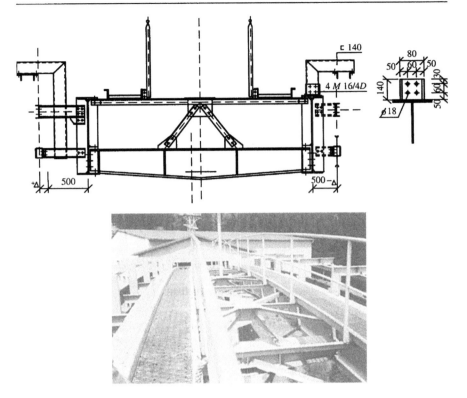

Figure 6.3   Lengthening of the original brackets

6.1.1.3   Refurbishment of a cable pipeline bridge while in operation [65]

The pipeline bridge across the river Labe near Neratovice was a prestressed spatial cable structure for the transport of sewage water. There were four load-bearing cables of 47.5 mm diameter fed through pylon saddles and anchored in gravitation anchor blocks. In the middle field with a span of 100 m were suspended triangular cross-frames carrying pipelines and a service footbridge (Figure 6.4). The rigidity of the structure was secured by reinforcing cables anchored in the foundation blocks of the pylons.

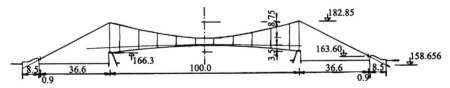

Figure 6.4   Schema of the pipeline bridge

Figure 6.5   Corrosion of load-bearing cables at the anchorage

In 1984 the load-bearing structure was in an emergency condition. Due to the high corrosiveness of the environment, the top layers of wires in the load-bearing cables (Figure 6.5) and a greater number of wires of the load-bearing suspensions (Figure 6.6) were damaged. Because of corrosion of a large part of the cross-section, the anchoring devices of the cables were considerably weakened (Figure 6.7). Other rigid parts of the structure (pylons, suspension frames) were not significantly damaged.

Figure 6.6   Corrosion of load-bearing suspensions

Figure 6.7 Corrosion of an anchoring device

Given the extent of the damage, it was necessary to repair the load-bearing structure as soon as possible. However, at least one pipeline of diameter 630 m had to remain in operation. In order to meet this requirement, the refurbishment proposed was designed so that the new parts of the load-bearing structure would as far as possible be independent of the original structure.

The proposal for reconstruction
Two pairs of new load-bearing cables of diameter 60 m were laid parallel to the old cables, and the original pylon saddles and load-bearing suspended cross-frames were modified. The original saddles for pairs of load-bearing cables were extended on both sides by means of two additional saddles of identical shape (Figure 6.8). The cables were connected to the saddle by pressure plates and completed stops on both sides of the pylon head, designed to bear a friction force of 200 kN. The suspended cross-frames on the load-bearing cables were divided so they were

(a)   (b)

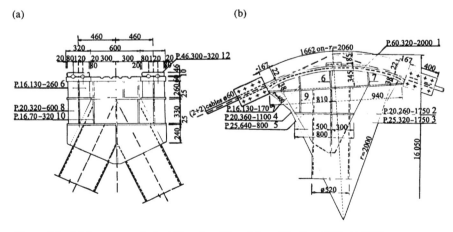

Figure 6.8   (a) Arrangement of additional saddles, (b) modification of the saddles

independent for each pair of cables. From the outer sides the top of the cross-frame a new anchoring device was fitted (Figure 6.9). The T-shape clips for the load-bearing cables were secured against lifting by bolted lower plates, and against shifting on the cables by individual stops designed to bear a friction force of 65 kN in the region where the suspension was longest.

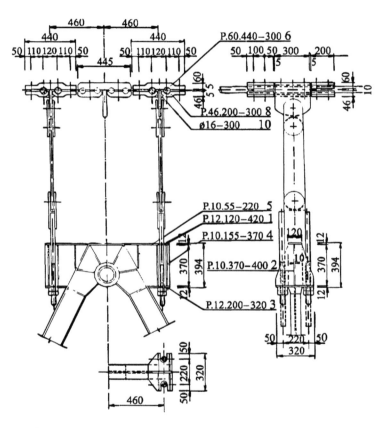

Figure 6.9 Proposed modification of the load-bearing suspensions and the upper part of the transverse frames

The new load-bearing cables were anchored by means of anchor binders placed in front of the anchor blocks under and close to the old load-bearing cables (Figure 6.10). The anchor binders were fixed to the rear walls of anchor blocks by means of draw bars aligned close to the side walls of the anchor blocks. The rear binder is fixed at the back of the anchor block via a reinforced concrete threshold, by means of which a vertical component of the force in the draw bar is transferred by friction to the anchor block (Figure 6.11).

(a)

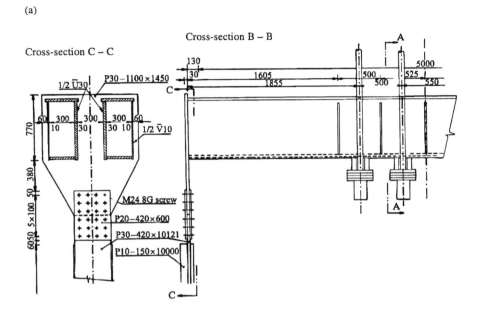

(b)

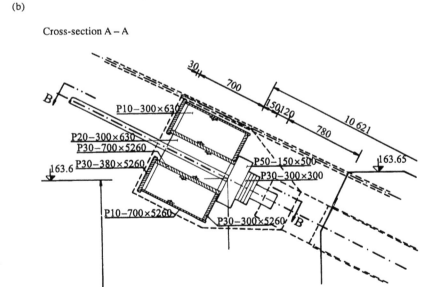

Figure 6.10 (a) Front anchor binder, (b) anchor binder with closed cross-section

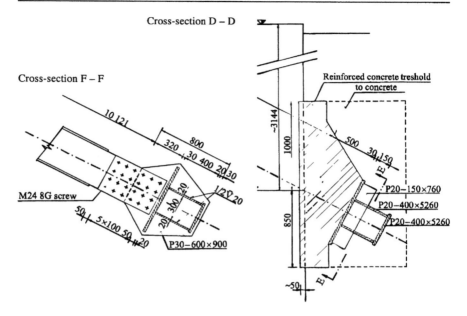

Cross-section D – D

Cross-section F – F

Reinforced concrete treshold
to concrete

M24 8G screw

P30–600×900

P20–150×760
P20–400×5260
P20–400×5260

Figure 6.11 Modification of the rear wall of an anchor block to support the rear binder

The course of reconstruction work

The construction works began by uncovering the anchor blocks to a depth of 5 m. There was a high level of underground water, which had to be drawn off during modification of the blocks. The repair of the steel structure began by modifying the pylon saddles, erecting the new transverse frame suspension devices and constructing the anchoring device near the anchor blocks. Meanwhile, the cables were

Figure 6.12 Setting the new load-bearing cables

being prepared. After being laid out on an adjacent passage way and marked out, their ends were fused together. The load-bearing cables were drawn along the roller track formed above the pylon saddles. When the cables had been set on the saddles, they were fixed by pressure plates and stops.

Before the new reinforcing cables had been erected, the original cables were loosened to about 20 per cent of the prestressing force by removing the bases under the sealing ends. Their anchorage could then be moved downward along the anchor grid and the new reinforcing cables lowered from above into their places. The new load-bearing and reinforcing cables were longer by 100 or 150 mm at each end, so they could be anchored using only a small force.

Moving the structure onto the new cables began with the reinforcing cables. For erection work over the river, a floating scaffold on a boat was used. The transverse frames were moved to the load-bearing cables in several stages, transferring the weight by gradual tightening of the bolts on the suspensions. After each stage the load-bearing cables were tightened. In order not to exceed the capacity of the clamping of the cables on the pylons, all four cables on one side of the bridge were tightened simultaneously while the old cables were gradually loosened.

During erection and moving, the geometrical shape of the structure was monitored, particularly the movement of the pylons in the bridge axis and slackness of the load-bearing cables between the pylons. Cable tightening was completed after the required geometrical shape of the structure with its axial forces had been reached in both the load-bearing and prestressing cables.

After completing this most complicated stage of the reconstruction, the old cables were removed, the footbridge was replaced and the anchor binders were supported. Then the draw bars connecting the anchor binders with rear binders on the anchor blocks were embedded in concrete in order to decrease the danger of cracks forming in the concrete that protected the draw bars against corrosion. The concrete was reinforced so that larger cracks could not develop. A view of the space prestressed cable structure after refurbishment is shown in Figure 6.13. Figure 6.14 shows the new anchoring device for the load-bearing cables.

Figure 6.13 The cable structure for the pipeline bridge after reconstruction

Figure 6.14 New anchor binders for the load-bearing cables

Great attention was paid to resisting corrosion of the structure. The anchor binders have closed cross-sections with smooth surfaces and outlets for rain water. All new steel structures were metallized with a Zn 80 + Al 120 layer and finished with a synthetic coating. The cables were protected against corrosion by a Resistin ML coating and two coatings of Resistin Car.

The sophisticated steel structure of this pipeline bridge was reconstructed in a relatively short time without interrupting operations. Approximately 25 tonnes of structural steel and 20 tonnes of cables were used in the work.

6.1.1.4 Reinforcement of the anchorage of load-bearing cables

Due to the incorrect design of a detail in the anchorage of the pipeline bridge across the river Váh at Piešťany [66], a slight slippage of the cables at their ends had occurred, causing some tilting of the pylons. This case involved a prestressed cable structure with a mid-field span of 75 m (Figure 6.15).

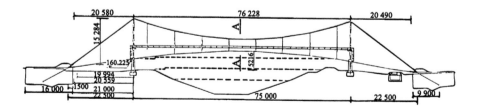

Figure 6.15 A schema for the cable pipeline bridge at Piešťany

During the assessing examination, after drain openings had been cleaned some water was poured off the anchor device (Figure 6.16). This permanently wet environment had accelerated the corrosion of the compound-filled cable ends.

Figure 6.16 Anchor device for the load-bearing cables before reconstruction

After examining the actual geometric shape of the cable system (Figure 6.17), a static calculation was carried out to check its status.

Figure 6.17 The structure of the pipeline bridge

On the basis of the results of both measurements and static calculation, it was proposed that the anchorage of the load-bearing cables be reinforced by means of cable clamping rings fitted to the back of the new pre-adjusted anchor binder (Figure 6.18).

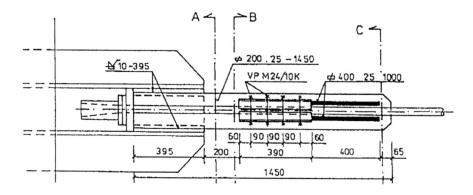

Figure 6.18 Reinforcement of the anchorage of the load-bearing cables using frictional clamping rings

The anchor device for the load-bearing cables after refurbishment is shown in Figure 6.19.

Figure 6.19 Anchorage of the load-bearing cables after reinforcement

6.1.1.5 Replacement of the prestressed cables of a footbridge

The steel load-bearing structure of the access footbridge across a branch of the Danube in Bratislava is three-field suspension system with spans of 7.0 + 72.0 + 7.0 m and V-shaped pylons (Figure 6.20). The reactive forces in the load-bearing and prestressed cables are absorbed by reinforced concrete anchor blocks.

Figure 6.20 Load-bearing cables and V-shaped pylon of the footbridge

The free width of the footbridge between rails is 1.3 m. The load-bearing suspensions are arranged in two inclined planes, and carry bearing binders on which the panels of the flooring and rails are located. The prestressed cables are fixed to the edges of the flooring and in longitudinal direction they follow the shape of the flooring. The load-bearing cables are fed through the support saddles of the inclined pylons.

During examination a slight overall slewing of the footbridge was discovered. This slewing could have originated in a faulty erection procedure, or from uneven prestressing forces in the stabilizing (prestressing) cables.

The surrounding tall vegetative cover prevented aeration, while organic matter together with rain water generated permanent high humidity that aggressively corroded the steel structure, especially the cables. In some places the vegetative cover was in the contact with the load-bearing structure, and some tree trunks and branches had penetrated the cable elements. The prestressing cables had been strongly corroded over all their length (Figure 6.21), though with uneven intensity.

Figure 6.21 A prestressing cable damaged by corrosion

To secure the operational reliability and prolong the life of the structure, the following modifications were made:

- The high vegetative cover from the surroundings of the pedestrian bridge was removed or treated
- After deep cleaning, the coating system on the steel and cable elements was renewed
- Prestressing cables that were heavily corroded were replaced (Figure 6.22).

After replacing the prestressing cables the re-alignment of the cable structure of the bridge was carried out.

Figure 6.22 A detail of the anchorage of the new prestressing cables

6.1.1.6 Modification of the anchorage of the load-bearing cables of a footbridge

A spatial prestressed suspension footbridge for pedestrians with spans of 25.0 + 100.0 + 25.0 m was built across the river Orava at Dolný Kubín. After 20 years of operation in a constantly wet environment, the sealed ends of two of the four load-bearing cables had been heavily corroded (Figure 6.23). The load-bearing cables were single-wound of closed structure, with diameter 50 mm.

After new cable ends were made and fitted, the cables were lengthened by threaded bars of diameter 120 mm (Figure 6.24, on the right).

Figure 6.23 The anchor block of the footbridge

Figure 6.24 Modification of the anchorage of the load-bearing cables

6.1.1.7 Rectification of geometric shape of cable system

The pipeline bridge in the new Winter Harbour in Bratislava was completed 20 years ago. It is a prestressed suspension structure with a 234 m span of wind stabilizing cables (Figure 6.25), by which at relatively small prestressing forces the required spatial rigidity of the structure is attained.

During the technical assessment of the bridge's condition, the geometric shape of the structure was measured (Figure 6.26). It was established that the current geometric shape differed from its shape on erection, because the slackness of the load-bearing cables had increased by 386 mm (Figure 6.27).

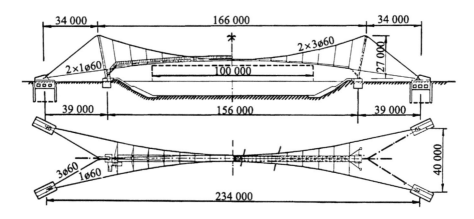

Figure 6.25 A schema for the spatial cable system of pipeline bridge in Bratislava

Figure 6.26 Measurement of the actual geometric shape of the structure

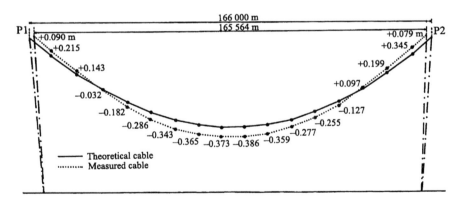

Figure 6.27 Results of the measurement of geometric shape

In this structure the load-bearing and prestressing cables were anchored in common anchor blocks, with the anchor devices accessible for inspection and maintenance.

On the basis of analysis of the measurement results and a checking static calculation, the geometry of the system was rectified by additional prestressing of the load-bearing and wind-stabilizing cables (Figure 6.28).

Figure 6.28 Anchorage of load-bearing and prestressing cables

6.1.1.8 Reconstruction of a footbridge

The original suspension footbridge across the river Váh in the village of Stankovany, Slovak Republic was built in 1955 with spans of 13.0 + 60.4 + 4.0 + 5.43 m.

(a)

(b)

Figure 6.29 The original structure of the suspension footbridge

When the physical life time of the load-bearing cables, suspensions and flooring had been used up, the original suspension system was replaced by a new system. However, the original ground constructions, anchor devices and pylons were reused. The method of branching the suspensions in the branch system is shown in Figure 6.30.

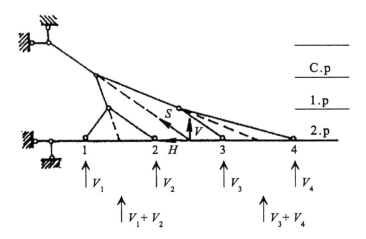

Figure 6.30 Method of branching the suspensions in the branching system

During reconstruction the original pylons were heightened by 3.2 m. In the new branching system (Figure 6.31) the reinforcing girder was suspended at angles on points in the middle span.

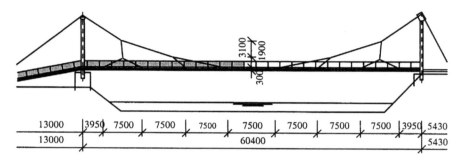

Figure 6.31 Schema for the new branching system

The inclined suspensions are made of tubes with diameters of 108.18, 83.14 and 83.9 mm. These tubes are connected at nodes by means of pins with diameters of 70, 55 and 40 mm. The details of the nodal structures are shown in Figure 6.32.

(a)   (b)

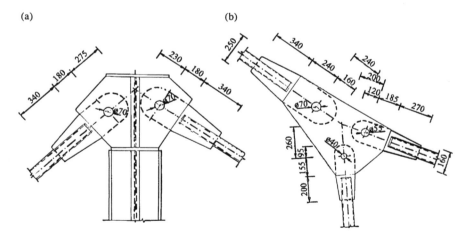

Figure 6.32 Details of the branching system: (a) pylon head, (b) branching node

The flooring was designed using longitudinal IPE 330 girders reinforced in select-ed places. There are binders at distances of 2.0 m and from U 120 (I 120). The flooring is formed by trapezoidal plates, and on these a concrete slab 40 mm thick is embedded above the upper edge of the plate and reinforced by grid. A view of the branching system after reconstruction is given in Figure 6.33.

Figure 6.33 A view of the branching structure

Given the structure of the damage and difficulty of repairing cable structures, some general principles that should not be neglected may now be listed:

1   The effect of the environment must be considered when choosing the constructional system and its details.
2   If a part of the structure is subject to harder conditions or has a significantly shoter life time than other parts, the possibility of its replacement should be considered early in the design of the project. A slight increase in initial investment costs can result in a considerable saving in financial costs and shorter duration of repairs, and can prolong the life of the structure as a whole.
3   Some disadvatages of cable systems, such as great deformability, shorter life of cable construction elements and decreased capacity of cables stressed by tension under twisting can be ameliorated if a new type of suspension system such as a branching system is used.

6.1.2   *Technical bridges*

6.1.2.1   Reconstruction of an access footbridge by prestressing

The access footbridge to the inlet tower of the water works at Mikšova on the river Váh, Slovak Republic was originally designed using coupled simple girders with a 36.0 m span. The load-bearing structure consisted of two steel main girders at an axial distance of 2.5 m apart, which were to have been coupled in the erection stage using reinforced concrete precast units. This coupling should have been carried out by means of a central erection support (Figure 6.34). However, after the main girders and precast units were put in place, due to the flooded, waterlogged and sloping river bed the erection support lost strength and ceased to function effectively, so the original plan for the coupling had to be abandoned. These events led to inaccessible deformation of the footbridge at this stage of construction. This deformation, as well as the need to be coupled with the girder structure, required the immediate reconstruction of the footbridge.

Figure 6.34 A view of the central erection support

The deformation of the main girders in the middle of span amounted to 142 mm. It was necessary to find the best way of providing the necessary raising so that the required coupling could be carried out, but due to many unfavourable factors (high water level, the size of the deformation of the footbridge, etc.) the possibilities were significantly limited. Four alternative solutions were considered (Figure 6.35).

The fourth alternative proved to be most advantageous (Figure 6.35(d)). This involved prestressing by means of a horizontal (direct) draw bar. This solution enabled the erection of a console of prestressing devices above the water line; see the schema for this solution in Figure 6.36. Each main girder is prestressed by two cables placed one above the other. The cables are made from 2×24 diameter P7 (2000 kN), and the prestressing force needed in the draw bar was 1900 kN.

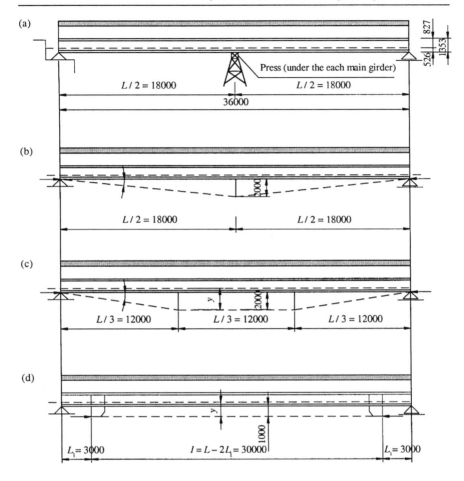

(a)

827
1353
526

Press (under the each main girder)

L / 2 = 18000   L / 2 = 18000

36000

(b)

L / 2 = 18000   L / 2 = 18000

(c)

L / 3 = 12000   L / 3 = 12000   L / 3 = 12000

y

2000

(d)

$L_1 = 3000$ $l = L - 2L_1 = 30000$ $L_1 = 3000$

y

1000

Figure 6.35   Schemas for the four variant reconstruction solutions

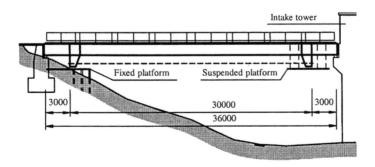

Intake tower

Fixed platform   Suspended platform

3000   30000   3000

36000

Figure 6.36   The schema for the reconstruction of the footbridge

The cables are anchored by means of friction anchors in the consoles. They are shown in Figure 6.37, together with the constructional adjustments to the main girders in the anchorage area.

Figure 6.37 A view of a connected console together with a prestressing device

Due to high shear stresses, the girder wall in the area of console is strengthened on both sides by diagonal reinforcement (made from angled section), so in this part of the structure there is a combination of lattice and web girders.

The working procedure was as follows. At the console connections, the transverse bracing of the walls was completed, along with the diagonal reinforcement. The transverse bracing of the main girders was adjusted and aeration of their lower chords was added. After the preparatory adjustment of areas with friction joints, the consoles were fixed to the lower chord of the girders by means of high-strength bolts from the platforms. When the necessary raising had been reached by prestressing, the steel girders were coupled using a reinforced concrete precast plate. An overall view of the footbridge after reconstruction showing the mounted prestressing draw bar is given in Figure 6.38.

Figure 6.38 A view of the footbridge after reconstruction

6.1.2.2 Reconstruction of conveyer bridges in a chemical plant

The original lattice steel structures of the NPK conveyer bridges were in permanent operation at the centre of a chemical plant where the environment had the highest degree of corrosive aggressiveness. During the inspection in 1978, workers scaled from bridge No.4b a layer of corrosive product of thickness 2.2 to 2.8 mm, and exceptionally up to 4.2 mm (Figure 6.39). To provide checking measurements of the material thickness, bridges Nos 1, 20 and 24 were chosen.

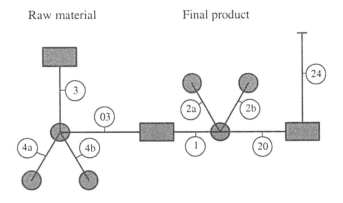

Figure 6.39 The arrangement of the NPK conveyer bridges

The greatest corrosive loss was found in bridge No.1 at an angle of 50.5° from the vertical, where the thickness of non-corroded metal of the flange was only 2.8 mm compared to the original 5.0 mm. Figure 6.40 shows a part of the original structures after dismantling in 1990. As well as extensive corrosion there was also found relatively large deformation (lateral deflection of the plane) of the verticals of the main girders on which the structure of lateral surface was fixed. The results of inspection and check measurements of material thicknesses showed the need to refurbish the steel structures of the conveyer bridges.

Figure 6.40 Corrosive damage to the structure

In order to preserve safe functioning of steel structures of the bridges, it was deemed necessary on the basis of static assessment and with regard to the corrosion losses to carry out the following measures:

- Reinforce all verticals of all bridges using IPE bending rigid profile (Figure 6.41)
- Reinforce the upper chord of bridge No.3 by 160.10 profile in the sections marked in Figure 6.41
- Add member D_4 made from two angle places at 56.5°, as in Figure 6.41, to the marked lattices of bridges No.2a and b, and No.4a and b
- Restore the coatings of all structures using a complete coating system
- Carry out a chemical analysis of the metal sample taken from bridge No.1 and test its microstructure to determine the changes in the mechanical and physical properties of the metal.

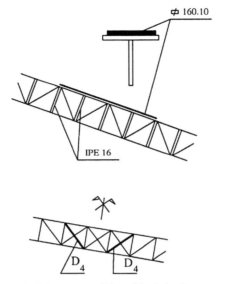

Figure 6.41 The proposed reinforcement of the original structures

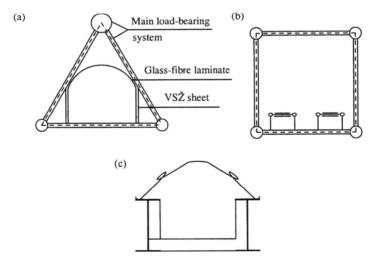

Figure 6.42 Proposed details: (a) lattice tripod girder, (b) quadripod main tubular girder, (c) web main girders

These measures prolonged the serviceability of the steel structures of the bridges and prevented them from failing. As the life of the bridges had been exceeded, it was recommended that a proposal for the gradual replacement of steel structures by new ones be prepared. Analysis of the technical condition of the steel structures attacked by corrosion revealed that in the original design of the steel structures the basic constructional measures for the prevention or restriction of corrosive process had not been taken. In addition it was shown that traditional types of conveyor bridges that use lattice structure are not suitable for areas where the environment has a high degree of corrosive aggressiveness.

The proposal for the new steel structures was based on the actual conditions under which the bridges would be used in the chemical plant [63]:

- The proposed construction must allow good access for grinding and restoration of the surface of the steel structure
- The possibility of using steel with an increased resistance to corrosion must be considered
- The best method of cleaning the metal surfaces before coating must be used, and also the best method of surface protection for the new structures.

The lattice structures composed of members of closed cross-section meet these requirements. The three-chord lattice structure (Figure 6.42(a)) can be considered to be a proper solution for the given conditions. The tubular lattice girder cross-section could also be a quadripod (Figure 6.42(b)). The covering of transparent glass-fibre laminate that replaced the windows can also be used to cover the conveyor belts. The sheet side covering (Figure 6.42(a)) having been corroded was replaced by a new one with an adequate surface protection. The load-bearing structure of the conveyor bridge was formed by two web main girders connected by binders, as in Figure 6.42(c).

When examining the possibility of using steel with higher resistance to atmospheric corrosion such as Atmofix, or Cortne it should be noted that in a very aggressive corrosive environment the conditions for creating a high-quality protective layer, will not be met.

The client required that the main girders of all bridges be welded web I girders. The semi-frames formed from these girders were to be covered with a glass-fibre laminate. The design solution for the bridges had to respect the existing conditions and due to connection to technical devices also to allow for the possibility of locating new bridges on the original structure. The flooring of all bridges was made of 50 mm wooden sheet piles placed on 140 × 160 mm cuboids. Wood can be an ideal material in such an aggressive environment. Given these requirements, the proposal for conveyor bridges Nos. 1, 2a, 2b, 3, 03, 4a, 4b, 20 and 24 was prepared. By 1990 all the bridges had been refurbished.

The axial distance between the main girders is 2400 and 4200 mm, the maximum span of the bridges is 33.16 m. On bridge No. 1, the original two-field bridge with one conveyor belt was replaced during reconstruction by a three-field web structure with spans of 38.45 + 17.45 + 27.00 = 82.90 m (Figure 6.43). In the transverse direction there is a semi-frame made from two main girders 1560 mm high and with I 200 binders. There are lattice tubular supports. The glass-fibre laminate semicircular shield has diameter 1600 mm. A view of new conveyor bridges is given in Figure 6.44.

Figure 6.43 Part of conveyor bridge No.1

Figure 6.44 The new conveyor bridges

In this example the optimum solution was sought for the structure of conveyor bridges that are in permanent service in a very corrosive environment. The design uses a combination of materials: web steel main girders, tubular stiffeners and supports, wooden flooring and a glass-fibre laminate shield to satisfy service reliability requirements, to prolong operational life and to reduce maintenance costs.

6.1.2.3 Repairs of gas pipeline bridging

The preliminary proposal for assessment of the steel structures of the pipeline bridging of the Transit Gas Pipeline (TGP) in Slovak territory was concerned with methods and apparatus. Complex assessing inspections of all bridging were carried out in 1998. Subsequently, the first repair and refurbishment work on the bridging was begun.

Assessing inspections
The aim of the assessing inspections was to determine the technical condition of bridging (pipes, pipeline bridges and non-load-bearing pipelines) and to gather documentation for an assessment book on each kind of bridging.

The pipeline bridges take the form of a lattice triple-hinged arch (Figure 6.45) with various spans and outspans at the crown. The structures are made from standard parts designed bridging sections of transit gas pipelines across rivers or streams. The main girder is built up from these parts in a doubled-up design.

Figure 6.45 The lattice three-hinged arch for the bridging of pipelines

Between the individual parts of the main girder there are doubled dutchman's trousers that by their trapezoid shape create the arch curvature of the main girder (Figure 6.46). Binders are located every 6000 mm, measured along the central line of the broken arch. They are bolted to the dutchman's trousers of the main doubled arches. The horizontal reinforcement between the doubled main arch girders is located in the plane in the middle of the main girders. The individual parts are lattice, tubular or all-welded. The pipes of the gas pipeline rest on bearings on the binders. These bearings consist of cylindrical rollers that enable the push fitting of the pipeline at different temperatures of pipeline and pipeline bridge.

Figure 6.46 A view of the girders and binders of the lattice bridging

Some pipelines are carried across rivers by non-load-bearing arches of various spans (Figure 6.47). At the flex points of the arch, the anchor bearings are fixed at an angle against anchor frames mounted on concrete foundations.

Figure 6.47 Non-load-bearing arch of a TGP pipeline

On the basis of the results of the input assessment in 1998 an assessment book was compiled for each kind of bridging. This contained the basic data on the bridging, its technical condition, any deficiencies detected, and proposed modifications needed to secure safety and operational reliability, and to prolong the life of the bridging.

Evaluation of results of inspections, proposed modifications and repairs
Pipeline bridges

- The pipelines are often carried by bridges of different height and they tend to approach one another more closely at the top of the arch (Figure 6.48).To secure and facilitate checking and maintenance of the pipeline over the whole length of the bridging, the pipelines separated are at the crown of the arch and the necessary gap between them is secured by spacing elements.

Figure 6.48 A view of the orientation and spacing of the pipelines on the pipeline bridge

- The pipeline rests against vertical stiffeners, mounted on rollers (Figure 6.49) set on the binders. On some bridges in places it rests on the upper chord of the foundation grill (Figure 6.50). Correct alignment of the pipeline is needed to prevent excess so that local stress at these places would not threaten the reliability of the whole pipeline.

Figure 6.49 Details of the stiffeners of   *Figure 6.50* Details of the binders of
               the rollers and bearing saddles   the foundation grill

- The pipeline does not always rest evenly on the cylindrical roller bearings (Figure 6.51); many placements and saddles may be non-functional (Figure 6.52).

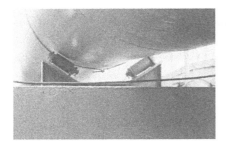

Figure 6.51 Lifting of a pipeline above the rollers of the bearing saddles

Figure 6.52 Slewing of the original bearing saddles and non-functional springs

Detailed analysis can reveal what is causing the 'lifting' of the pipeline above the bearing saddles of arched supporting bridge structures. On many bridges, due to uneven deformation of the pipeline or the bridge, some supporting saddles may be non-functional (lifting of pipeline over the arched bridge) or the pipeline may be slewed so that the springs on the bearers are compressed more tightly on one side, compared to the other.

The results of analysis showed the original expectation that the pipeline would be fully carried by pipeline bridges had not been realised, and that the pipeline bridges cannot be made to operate properly using the original bearing saddles. In the interactions between the bridge and the pipeline, the influence of the pipeline before and after the bridge, in the ground structures (brake and anchor blocks of the pipeline), and differences in climatic and operational temperatures are all significant. The original bearings are being gradually replaced, and new designs for these have been proposed (Figure 6.53).

Figure 6.53 Older type of lower telescopes

Due to continuing problems with the details of bearings that use cylindrical rollers a new method was proposed – a suspension system for the pipeline. The results of a theoretical analysis of the actual operation of pipeline bridges in interaction with the pipelines they carry confirmed that the pipeline can be suspended at a distance of 12.0 m. The constructional solution for the suspension for Transit Gas Pipelines I a, Transit Gas Pipelines II and Js 1200 is shown in Figure 6.54. For alignment of the pipeline position in the vertical direction, springs with a capacity

Figure 6.54 The suspension system

Figure 6.55 A view of the external vertical columns with upper telescopes

of at least 120 kN were designed. The suspension draw bars are made of circular bars with threads at the free end. The marginal supporting brackets are connected to the upper chords of the main girders of pipeline bridge (Figure 6.55); internal vertical brackets are fitted on the original binders of the bridge. As well as the system of pipeline, a new type of saddle with lower telescopes and sufficient compression capacity in the springs were devised (Figure 6.56). These tools are fitted to the original bridge binders (Figures 6.57 and 6.58).

Figure 6.56 The vertical columns of the suspension system (external and central)

Figure 6.57 New lower telescopes

Figure 6.58 A view of the fitting of new saddles to the binders

- The constructional details of the main girders, the connection of the main girders to the supporting binder (Figure 6.59), the horizontal stiffeners for dutchman's trousers (6.60), the connection of longitudinal I girders to verticals (Figure 6.61) created undrained receptacles. To ensure the drainage of these important details of the structure, the openings were sealed off and drainage holes for collected water were cut.

Figure 6.59 A detail of the connection of the main girders to the supporting binder

Figure 6.60 A view of the dutchman's trousers of the main girders

Figure 6.61   A detail of the connection of the internal girder to the verticals of the main girders

- On the main girders of pipeline bridges, notably on verticals (Figure 6.62) and diagonals, there were longitudinal cracks and often also local openings in the cross-section of the pipes. These pipes were constructed with longitudinal welds, which should not have been used in bridge structures. The cracks in the diagonals, verticals and welded connections were all repaired. In cases where the pipes could not be hermetically sealed by means of welds, the broken pipes were replaced (Figure 6.63). If while cutting out the broken part the opening in the pipes runs from node to node, the repair is done by cutting off the broken verticals and diagonals along the whole length and replacing then with a new member. Damaged pipe members in the binders were repaired in the same way.

Figure 6.62   A cracked vertical on a main girder

Figure 6.63 Replacement of a damaged vertical

- Where there is local or global buckling of some load-bearing elements, the condition of the diagonals in the horizontal reinforcement is monitored and if there is considerable increase in the deformation (Figure 6.64), the buckled members are replaced. The influence of deformation and weakening of the upper table of the binder is assessed, and on the basis of the results of this assessment, the table may be reinforced (Figure 6.65).

Figure 6.64 A buckled diagonal in a wind stiffener

Figure 6.65 Reinforcement of the upper tables of a binder

- In the chords of the foundation grill (Figure 6.66), the drainage openings and cuts may not be functional. The bearing plates of the foundation grid on the support may not reach the concrete walls, and the anchor bolts are often deformed (Figure 6.67). The concreting of the bearing plates needs to be repaired and anchor openings are inserted in the concrete faces. The bearing plates of the distribution binders near their location on the reinforced concrete foundation structure are sealed by pouring.

Figure 6.66 Corrosion damage to the binders of the foundation grills

Figure 6.67 Deformed anchor bolts

Non-load-bearing pipelines

The horizontal girders of anchor frames may not be drained sufficiently, and the bearing plates of vertical anchor girders may not bear directly onto the concrete faces (Figures 6.68 and 6.69). The lower clamping rings of the pipelines fitted with stops consisting of thin plates, but these stops may not sit symmetrically against the clamping rings. Any faults discovered can be repaired as with lattice bridging.

Figure 6.68 A view of the anchorage of a non-load-bearing pipeline

Figure 6.69 Another view of the anchorage of a non-load-bearing pipeline

6.1.3 Reconstruction of the Mária Valéria Bridge

The bridge across the Danube between Štúrovo and Esztergom, named after Mária Valéria, the daughter of Emperor Franz Joseph, was built in 1895. The three internal spans of the steel structure were destroyed in the Second World War. Only the outer spans of the upper part, river piers and bridge supports (Figure 6.70) withstood the war. This was the only road bridge along the whole length of the Danube that had not been reconstructed in the following half century. People and cars were transported seasonally by ferry.

Figure 6.70 The condition of the bridge before reconstruction

On 16 September 1999 the Prime Ministers of the Slovak Republic and Hungary signed the inter-governmental agreement for the reconstruction of the bridge. Since 11 October 2001, after 57 years, the frontier crossing across the river has been open to pedestrians, cyclists, cars, lorries (up to 3.5 tonnes), and coaches.

Expressed in modern terms and conceptions, the project involved a lattice girder bridge with a simple bedding of individual fields. The division of fields of the original bridge is harmonic; the main girders had spans of 83.5 + 102.0 + 119.0 + 102.0 + 83.5 m (Figure 6.71).

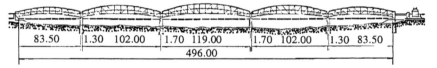

Figure 6.71   A schema for the original bridge structure

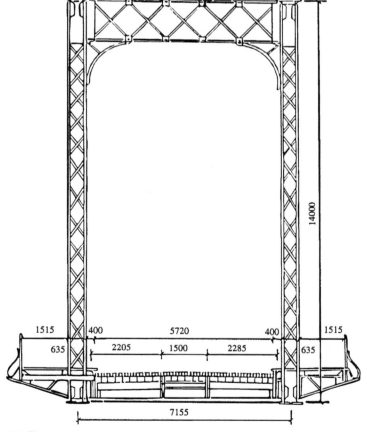

Figure 6.72   Transverse arrangement of the original bridge

At the time of its construction the middle span of 119.0 m was the largest simple-bedded bridge span built across the Danube. The theoretical height of the crescent-shaped curved chord main girders in the middle field is 14.0 m, in the adjacent fields 12.0 m and in the outer fields 10.0 m. The lower chord of the middle field was curved so that over a length of 50.0 m (the width of the channel opening) its lower edge was 6.9 m above the maximum river level. The supports of the main girders were located on one level so that the longitudinal line then was horizontal. The profile of the gradient line was polygonal, decreasing from the centre of the middle span by 0.2 per cent, in the adjacent by 0.6 per cent and in outer spans by 1 per cent. The steel structure of the bridge was dimensioned to bear the impact of railway transport, which was shown by the constructional solution used in the floor skeleton (Figure 6.72). The internal and outer ledgers of span 8.5 m and 8.35 m were riveted onto the binders. The axial distance between internal binders was 1.5 m. The original wooden floor was later replaced by a reinforced concrete slab.

Alternative proposals for the reconstruction of the bridge
The possibilities of partially reconstructing the damaged bridge were assessed in a study [67]. The aim of this study was to devise reconstruction proposal that would use the existing pillars and load-bearing structure in the outer fields. On the basis of results of the assessing examination, material tests and checking static calculations, utilization of other parts of the existing steel structure of the bridge in the outer spans would be considered. In the proposal it was necessary to respect the prominent location of the bridge in relation to the protected historical monument, the Esztergom Basilica.

In the studies [67] and [68] four alternative solutions for reconstruction were put forward. The first alternative proposed reconstructing the three missing internal fields of the bridge in the original (historically revered) shape. The next two solutions proposed a coupled lattice, and three-span web girder. In the fourth alternative, a three-span symmetric suspension system would represent this modern style of bridge building, with the original outer spans as a reminder of the war. After their later replacement by web girders, the bridge would then have a uniformly modern style.

After the specification of some marginal conditions had become more precise, another study [69] proposed the reconstruction of the steel structure in all fields. This would involve a five-field continuous lattice girder bridge with orthotropic floor. Reconstruction of all five fields with a minimum road width of 6.5 m had the advantage that the problem of the different life time of the outer (original) and the new internal fields of the bridge would be eliminated [70].

The adopted reconstruction proposal

After evaluation of all the alternative solutions with respect to economy and the prominent location of the bridge particularly in relation to the Esztergom Basilica, the parties involved agreed to reconstruct the bridge in the original shape, which would harmonize with the surroundings while using the existing parts of the load-bearing structure of the bridge in the outer fields, land abutments and river piers. Reconstruction to the original traffic parameters would at the same time eliminate heavy traffic from the bridge.

In the resultant proposal the bridge consists of a main structure with five fields of spans 83.5 + 102.0 + 119.0 + 102.0 + 83.5 m and a short field on the Esztergom side of span 16.2 m (Figure 6.73). The line of the bridge is higher in comparison to the original due to the required gradient dimensions of the new channel openings. In reconstruction the original 1895 structure in the outer fields are used, with the necessary modifications. For the three internal fields in which the original structure was destroyed, the new structures proposed have profiles identical to the original bridge.

On the main bridge an orthotropic floor adopted covered by a 90 mm thick asphalt road was proposed. The plate sheet of the floor is 12 mm thick. The floor binders are located at the nodes of the main lattice girders. The floor has three longitudinal stiffeners, ledgers with an axial distance of 2500 mm. The floor plate sheet is reinforced not only by binders and longitudinal stiffeners but also by transverse stiffeners of trapezoidal shape. The use of transverse stiffeners on the orthotropic floor is contrary to normal practice but has the advantage that it avoids the need for the complicated connection of a large number of closed cross-section stiffeners. Also on the main bridge are pavements located on the outer sides of the main girders, with a width of 2250 mm. Their floor consists of a reinforced steel plate 10 mm thick covered by an antislip layer. On the edges the flooring plate is bedded onto the longitudinal pavement girders with axial distance of 2250 mm.

The longitudinal pavement girders are supported at nodes of the main girders by lattice brackets. The outer edging longitudinal girder is of welded I cross-section. In order to preserve the style of the original bridge, the insertions in the walls are shaped so as to give the impression of a lattice girder. The inner filling of the rail is designed as latticing similar to that used on the original bridge. Some details of the pavement are shown in Figure 6.74.

In the outer fields the bridge has curved chord lattice main girders of a crescent shape with a span of 83.5 m. The axial distance between the main girders is 7.15 m, their height at bedding is 6.45 m, and in the middle of span it is 10.0 m. The cross-sections of the members are riveted. These chord members are double-sided with open cross-section, and the cross-sections of web members are modified to fit them. The diagonals and verticals are constructed as articulated members with connecting elements of flat steel. In the cross-section of the two outer lattices, the bridge

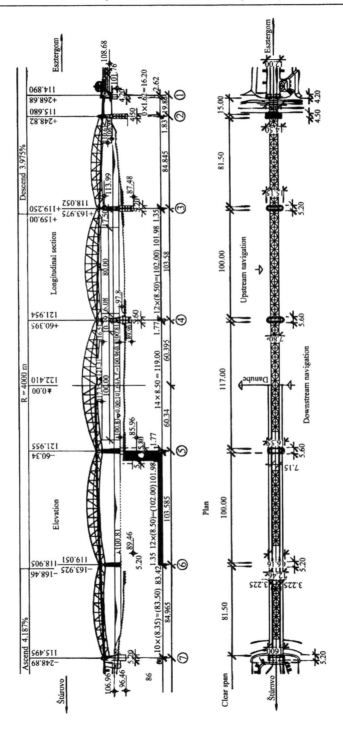

Figure 6.73 Schema for the bridge in the proposal adopted

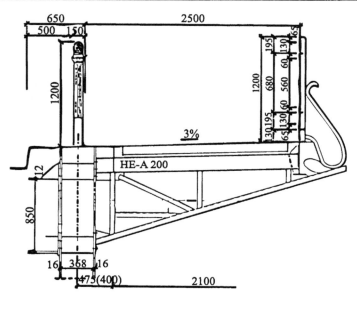

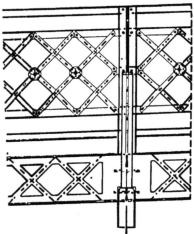

Figure 6.74 Details of the pavement

is arranged openly. During reconstruction, the main girders, upper and lower wind stiffeners and upper parts of the transverse stiffeners from the original structure are all used. The cross-section of the outer fields of the bridge after modification is shown in Figure 6.75.

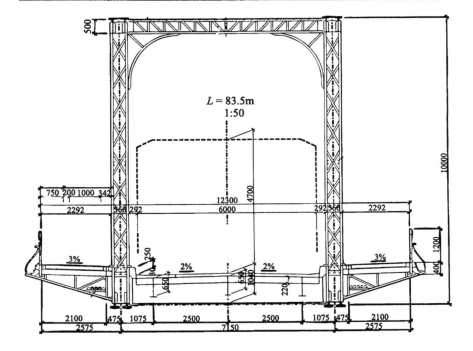

Figure 6.75 A cross-section of the outer field of the bridge

To ensure the static strength of the outer fields it was necessary to reinforce:

- The cross-sections of the verticals of the main girders (Figure 6.76), minimum in their sector with full wall
- The cross-section of the upper chord in the first lattices
- The cross-section of the upper chord damaged during the war (Figures 6.77 and 6.78)
- All diagonals of the upper longitudinal stiffener: the original cross sections were not suitable so they needed to be replaced by cross lattice member of 2 L 100.12

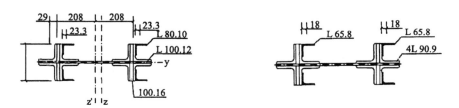

Figure 6.76 Examples of the reinforcement of the vertical main girders

- The lower chord of the end transverse stiffeners by adding a table of 250×10
- The end transverse stiffeners completed under the table by diagonals of 2 1 100×12
- Reduce the buckling length of the verticals of the lower longitudinal stiffener in the middle of the field and at the adjacent nodes by suspending the floor over half their length onto the lower chord of the binder.

To enable connection of the pavement brackets it was necessary to modify the reinforcements of verticals 3 and 4.

During Second World War the upper chord of the main girder near the central vertical on the Bratislava side was damaged. The flange angle and about one quarter of the width of the table was also damaged. At this node the diagonal of the upper longitudinal stiffener was also torn off and had to be replaced (Figure 6.77).

Figure 6.77 Local damage to the upper chord of the main girder

The damaged upper chord was reinforced by an additional table P. 20 – 650 × 1260. The base of 20 mm thickness under the reinforcing table was perforated in some places by rivet heads (Figure 6.78).

The structure of the floor is the same in all fields with transversally arranged closed reinforcements (Figure 6.79). The new steel structures are of the same shape in both internal and external fields. The axial distance between the main girders is 7.3 m.

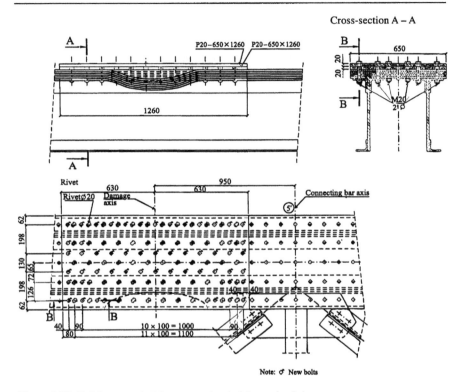

Figure 6.78 Reinforcement of the upper chord of the main girder

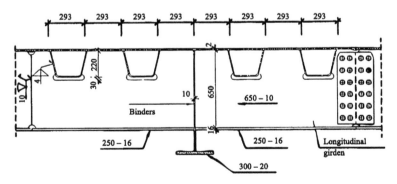

Figure 6.79 A detail of the orthotropic floor

The span of the main girders in fields II and IV is 12 × 8.5 m, i.e. 102 m. Their height above the bedding is 6.45 m, and in the middle of the span it is 12.0 m. In the middle field III the structure has also curved chord lattice main girders of crescent shape with a span of 14 × 8.5 m = 119 m. The total height of the main girders above the bedding is 6.45 m, and in the middle of span it is 14 m. The cross-sections of the members of the main girders in these fields are welded after pre-assembly by means of friction joints at the manufacturers. The chord members closed cross-section, the web members are of I cross-section, while there are incisions in the wall that cause the web members to act statically as an articulated lattice. The end verticals have a reinforced closed cross-section. These verticals are stressed by an axial compressioned force and their buckling length in the plane of the main girders has been reduced by stabilizing members, located at mid-height of the main girders. These stabilizing members are connected at their ends to the reinforced end verticals. The outer lattices of the bridge have an open cross-section. Figure 6.80 show a typical cross-section of the new fields of the bridge, while Figure 6.81 shows a typical node in the main girders of the new fields.

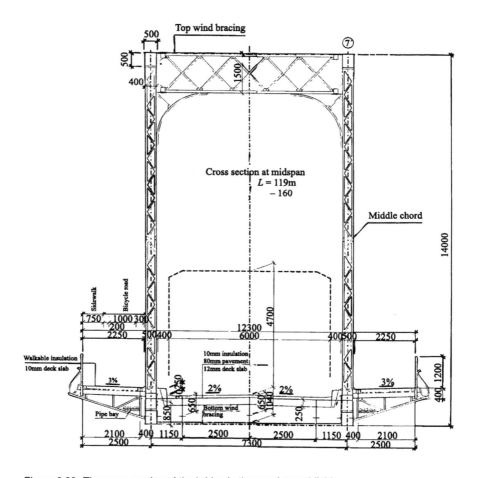

Figure 6.80 The cross-section of the bridge in the new internal fields

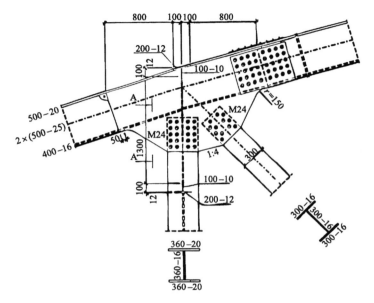

Figure 6.81 A node of the main girders in the new fields

The details of the new fields are in accordance with the style of the original bridge. The floor, the pavements, the upper and lower longitudinal stiffeners, as well as the transverse stiffeners, are all designed using steel of strength class 37 (S235). The main girders use steel of strength class 52 (S355).

Manufacture and erection of the steel structure of the bridge
The documentation of the proposal for tender for the reconstruction of the bridge was prepared by Pont-TERV Budapest and Dopravoprojekt, A. S., Bratislava. The detailed static calculations for the steel load-bearing structure [72] and the proposal for reinforcing the original parts in the outer fields [73] were prepared by members of the Department of Metal and Wooden Structures of Faculty of Civil Engineering at the Slovak Technical University in Bratislava. Drawing documentation was prepared in co-operation with Dopravoprojekt, and Steel OK, Levice.

In the middle of October 2000 the work began. Hungarian companies GANZ and Közgép won the tender for construction of the bridge, and to do so established a consortium under the name GANZIS.

The reconstruction work on the external supports Nos.1 and 7 and river piers No.2 to No.6 were done by Inžinierske stavby, Košice. This involved repairing and raising the supports and pillars needed due to the change in the gradient of the bridge. Közgép Rt. Budapest prepared the refurbishment and reinforcement of the steel structure of the two existing outer fields and replacement of the short bridge of 16 m span on the Hungarian side.

The load-bearing elements of the three internal fields were made in the bridge building factory of Ganz Acélszerkezet Rt., Budapest. These elements were transported on the Danube to AssiDomän in Štúrovo in three stages from April to June 2001. Here, on the erection platform the main girders were first dismantled into three parts and then by means of a floating crane were reloaded in vertical position onto coupled barges. The whole steel structure of the bridge apart from the pavements, was gradually put together in this way (Figure 6.82). The erection of one bridge field took approximately three weeks. The individual bridge fields of 520 and 610 tonnes weights were transferred to the coupled barges, which were equipped with lifting devices and were transported to the bridge axis (Figure 6.83). Details of the construction of the bridge are shown in Figures 6.84 and 6.85. The total consumption of constructional steel and reinforced steel was 2100 tonnes and 300 m³.

Figure 6.82 Erection of the main girders

Figure 6.83 Floating and erection of the bridge fields

Figure 6.84 Some details of the reconstructed bridge

Figure 6.85 A view of the reconstructed bridge

6.1.4   Special structures

6.1.4.1   Reinforcement of the gate of a lock chamber

After the failure of the left-hand part of the lock chamber (see Section 2.3.2), it was necessary to assess the usability of the upper gate of the lock chamber in the Gabčíkovo water works.

To increase capacity and operational reliability, and to secure a temporary usability, the steel structure of both parts of the lower gate have been reinforced. This involved the direct reinforcement of vertical girders Nos.1 and 2 by additional tables and walls, which were connected to the original structure by high-strength M24/10K bolts (Figure 6.86). As the requirements of Czechoslovak standard 73 1495 for friction connection could not be met in reinforcing the steel structure of the gate the capacity of the M24/10K bolts was specified for the case of thick connection.

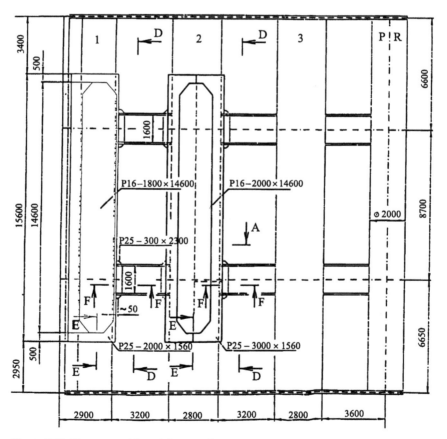

Figure 6.86 The proposal for reinforcement

The proposal for reinforcing the steel structure of the gates (Figure 6.87) was developed from the analysis of physical condition of the structure:

- As on the failed part, cracks were also found in the same welds in the right-hand part of the lock chamber. There was a crack approximately 1.0 m long in the first vertical girder in the 16 mm thick inclined wall, of the same kind as in girder No.1 of the failed part.
- The stresses in the original structure at maximum load are high; in some places they exceed the calculated values of the strengths of the steels used. The gate is made of 15 422.5 steel in the exposed parts which offers little resistance to propagation of brittle cracking. The welded were sensitive to cuts.

On the construction site, without higher pre-heating and heat it was not possible to make sufficiently reliable weld joints.

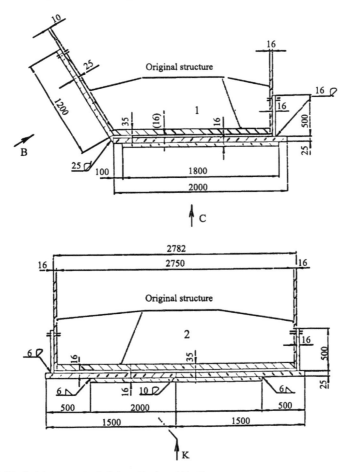

Figure 6.87 Reinforcement of girders No.1 and No.2

When assessing the reinforcement needed, the following factors were taken into consideration:

- Additional modification (reinforcement) of the steel structure, part of whose life has already been exhausted, can reduce the stresses to the prescribed levels, and at loadings below minimum operational level, can temporarily secure sufficient capacity in the structure.
- Though such reinforcement of parts of the gate can prevent the development of new cracks and the propagation of existing cracks already in the structure only partially not completely.
- The capacity and the reliability of the reinforced structure depends on those being sufficient capacity in the original welds (especially the neck welds) that have not been reinforced (that is possible only by use of additional neck angles bolted to the original elements).
- The additional reinforcing elements and parts were made on site, this work was done under very difficult conditions, partially under water.
- Given these factors, for the immediate short-term operation of the highly exposed steel structure of the reinforced gate parts, a procedural document for the temporary use of the modified structure of the lower gate was prepared. After a few months use of the reinforced structure this was replaced.

6.1.4.2 Raising a cable bridge in a water works

A reinforced-concrete structure of a cable channel 9.87 m long between a water power plant and a dam in forms the supporting system for the rails of the runway for a 20/3.2 + 3.2 tonne portal needle crane. During operation the cable channel settled by 240 – 250 mm. The necessary rectification of the level of the crane runway was done by raising the cable channel using additional girder-style steel girders bedded on the massive reinforced concrete structure of water power plant itself. The loading was transferred to these girders by means of five coupled draw bars and underfitted binders (Figure 6.88). The cross-section of the main girders (Figure 6.89) is variable, their span at the stage of elevation is 12.68 m. The basic elements of the cross-section consisted of rolled 2 I 500 mm, and welded to these elements from the inside were HR – 12 mm, forming the upper chord of a closed cross-section. The vertical draw bars were made from ϕ = 95 mm bars, and the under-fitted binders were of 2 I 500 – 7000 cross-section (Figure 6.89).

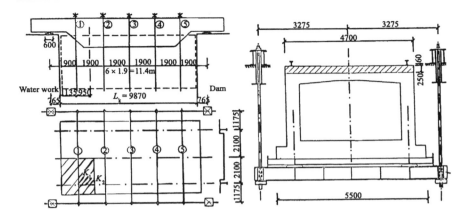

Figure 6.88 Schema for the additional steel girders

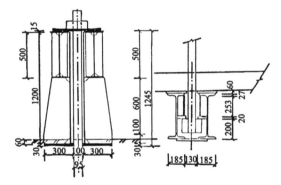

Figure 6.89 Cross-sections of the main girders and under-fitted binders

Figure 6.90 Raising the cable channel

The cable channel was raised using four 100 tonne hydraulic presses and after loading tests the rails of crane runway were aligned (Figure 6.90).

6.2 Engineering structures

6.2.1 Assessment of the technical condition and repairs to cylindrical tanks

6.2.1.1 Characteristic of internal corrosion damages in tanks for liquid fuels

Corrosion in a steel tank for liquid fuels takes different forms not only in the main structural elements of the tank, e.g. the bottom, the shell, the roof, but also in different parts of these elements. This is due to the very diverse corrosive environment that exists in the interior of a tank with a fixed roof. Four zones may be distinguished:

1 *The gaseous zone* This includes the roof structure and the top part of the shell filled with vapour saturated with hydrocarbons mixed with air. This mixture is characterized by variable humidity and various concentrations of the corrosion-activating agents, such as H_2S, O_2 and CO_2 depending on their concentration in the air in the vicinity of the tank. The steel structural elements in this zone of the tank are above the liquid fuel level for the whole operating period of the tank.

2 *The tank capacity zone active while the tank is in operation* This includes the surface area of the shell that is temporarily affected by the liquid fuel (when the tank is full), or a mixture of the fuel vapour and air (when the tank is empty). Moreover, if in course of emptying the tank its interior is filled with a large quantity of humid air, the sheet shell in this zone may be subject to the action of an emulsion of hydrocarbons in water containing much dissolved H_2S and oxygen.

3 *The 'dead' capacity zone in the tank* This is the area below the suction outlet level. Under operational conditions it is always filled with liquid fuel and water below a layer oil separated from the hydrocarbons by gravitation. The water contains various corrosive agents, often including polluted sulphur compounds, sodium chloride and anaerobic bacteria reducing sulphates. Corrosive aggressiveness in this zone depends on the frequency and effective drainage of water collected under the oil layer and on the frequency of external refilling.

4 *The surface of the bottom on the tank* Here there is an accumulation of oil residues and sediments, as well as corrosive by products that have fallen from the roof, such as various sulphides and oxides.

The boundary between the two first zones is not stable but oscillates over a very wide range. This depends on the operational circumstances, in particular on the frequency of filling and emptying the tank.

In the interior of a tank with a floating roof it is possible to distinguish similar corrosion zones with the exception of the gaseous one, which is precluded by the nature of the tank. On the other hand, in the 'dead capacity' zone, the water that collects beneath the oil layer will be significantly replenished by rain water flowing down the shell through leaks in sealing of the floating roof.

Corrosive damage to the bottom and in the dead zone of the tank present the biggest threat to the tank's leak-tightness. Damage here mainly consists of pitting corrosion, which is intensified by the oil residues and sediment spread over the bottom and up to a certain height of the shell, as well as deposits of sulphides and oxides falling from the tank's roof as it corrodes. Exceptionally favourable conditions for the development of corrosive micro galvanic cells are created under the sediments, where pockets form with a variable supply of oxygen from the water beneath the oil layer. Locations with oxygen insufficiency become increasingly anodic which initiates rapid development of corrosive pitting in areas at the surface of the sheet steel that are non-homogeneous, such as the steel grains at the borders, where there are non-metallic inclusions (Figure 6.91), where there are structural disturbances caused by welding (Figure 6.92), and where there has been heterogeneous plastic working. Such anodes of very small surface area concentrate the corrosive micro-cell current sulphides and hydrogen sulphide in then water beneath the oil layer, which serves as the cell's electrolyte, significantly speeds up the corrosive pitting process. The chlorine ions, as well as the hydrogen sulphide, penetrate into the thin passive layer of the steel, weakening its protective proper-

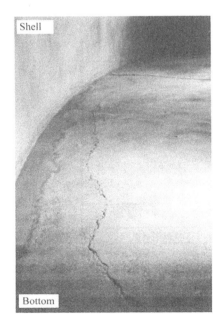

Figure 6.91 Corrosion pitting in areas of non-metallic inclusions

Figure 6.92 A corrosion pit near the fillet weld connecting the shell to the bottom inside a tank used for storing gasoline

ties, and by accumulating in the corrosion pits they intensify the aggressiveness of the domical solution there and rapidly speed up the deepening of the pits. The rate of pit deepening, both in the bottom and the bottom zone adjacent to the tank shell, can be 2 – 3 times faster than in the main part of the shell.

The corrosion rate is influenced by the anaerobic bacterial reducing sulphides. The distribution of bacteria in a fuel oil tank of 4000 m³ capacity was described by Wright and Hostetler (Figure 6.93) [77]. Development of micro-organisms in the tank was closely connected to the presence of water accumulated in the bottom zone under the stored oil product. At the water—fuel boundary phase there are chemical processes that lead to the formation of so-called bio-films and the creation of favourable conditions for the fast growth of micro-organisms. Unless their growth is prevented or limited, contaminatants will eventually be deposited and corrosive processes will begin. These corrosive processes involve metabolic products of the microorganisms (biogenic acids and sulphides), microbiologic absorption of oxygen, and cathode depolarization caused by the hydrogenase enzymes. Micro-organisms may cause the biodegradation of the protective steel shells, which are not always resistant to the microbiologic decomposition. Further, they lower the quality of the liquid fuel stored in the tank. Under exceptionally unfavourable conditions of use of tanks for liquid fuels, the corrosion rate in the bottom zone may be 0.7 – 1.0 mm per year.

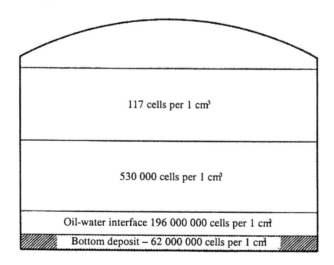

Figure 6.93 Quantity of micro-organisms in the fuel oil in particular tank shells [1]

Table 1 sets out the depths of the corrosion pits found in tanks with floating roofs [2]. These tanks were used for long-term storage of oil along the main pipeline in Poland.

Table 6.1 Depths of corrosion pits in oil storage tanks

Tank capacity [m³]	Uninterrupted operational use [years]	Sheet steel thickness according to the project [mm]		Greatest corrosion pit depth [mm]	
		Bottom perimeter and mid-height	Low shell ring	Bottom	In shell up to 300 mm above bottom
12000	30	9 and 7	11	2.0	10.0
12000	30	9 and 7	11	1.5	7.0
12000	30	9 and 7	11	3.0	5.0
32000	20	13 and 8	19	3.8	8.0
32000	20	13 and 8	18	2.5	5.0
50000	20	17 and 8	24	1.5	6.0

None of the internal bottom and shell surfaces of the tanks listed in Table 1 were protected against corrosion during their construction or operating period. The corrosion damage detected in the low shell area was particularly severe. The condition of the first 12000 m³ capacity tank was determined first both on account of the depth of the corrosion pits and the intensity of their occurrence (Figure 6.94).

Figure 6.94 Corrosion pits in a tank filled with crude oil after 20 years of operation (the plates were not protected against corrosion)

6.2.1.2 Determination of corrosive damage to the bottom, the shell and the roof

Estimation of the corrosive damage to the bottom of a tank presents a particular problem because inspection is only possible on the internal side of the tank, whereas corrosion of the bottom proceeds on both its sides. There is a general opinion that the corrosive process on the foundation side is insignificant, provided that the external layer of the sand foundation is well-saturated with heating oil. This conviction is justified, but only in the case when the tank bottom adheres closely to the sand foundation. However, along the bottom perimeter local lack of contact between the bottom of the foundation is likely to occur and in such places corrosive process can be more evident, as a result of the effect of a wet air on the sheet steel.

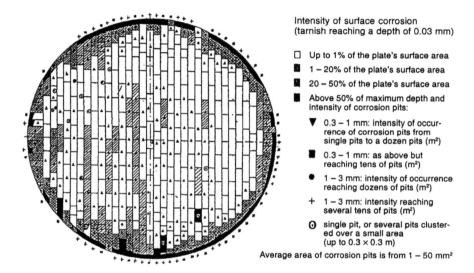

Intensity of surface corrosion
(tarnish reaching a depth of 0.03 mm)

☐ Up to 1% of the plate's surface area

▨ 1 – 20% of the plate's surface area

▨ 20 – 50% of the plate's surface area

■ Above 50% of maximum depth and intensity of corrosion pits:

▼ 0.3 – 1 mm: intensity of occurrence of corrosion pits from single pits to a dozen pits (m²)

▮ 0.3 – 1 mm: as above but reaching tens of pits (m²)

● 1 – 3 mm: intensity of occurrence reaching dozens of pits (m²)

+ 1 – 3 mm: intensity reaching several tens of pits (m²)

⊙ single pit, or several pits clustered over a small area (up to 0.3 × 0.3 m)

Average area of corrosion pits is from 1 – 50 mm²

Figure 6.95 Corrosion damage in the tank bottom, from the sand foundation side (the tank was used for 23 years)

Some interesting illustrations of corrosive wear on both sides of the bottom was collected during the disassembly of a damaged tank of 32000 m³ capacity, after 23 years of operation [79]. Figure 6.95 shows the corrosive damage to the bottom viewed from the foundation. Substantial losses in the sheet steel thickness were found only around the perimeter, while in the centre the intensity of pitting corrosion was insignificant, with the depth of the corrosion pits ranging from 0.3 to 1.0 mm. Much heavier corrosion losses were observed in the bottom sheets on the side of the tank (Figure 6.96). In numerous locations the corrosion pits reached a depth of approximately 6.0 mm, and these were spread over much of the surface of the sheet steel. It is worth pointing out that the tank bottom was made of sheet steel 8.0 mm thick in the centre, and 13.0 mm thick at the peripheral. The high rate of corrosion is related to the extremely aggressive corroding agents that occur in the 'dead zone' of the tank, such as below the suction flanges, where the water beneath the oil layer is, as a rule, very polluted with sulphur compounds and sodium chloride, and moreover, with some strongly aggressive residue precipitates from the liquid fuel, in particular from crude oil, that settle on the bottom.

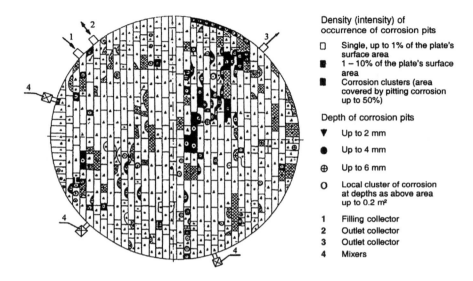

Density (intensity) of
occurrence of corrosion pits

☐   Single, up to 1% of the plate's
    surface area
▨   1 – 10% of the plate's surface
    area
■   Corrosion clusters (area
    covered by pitting corrosion
    up to 50%)

Depth of corrosion pits

▼   Up to 2 mm
●   Up to 4 mm
⊕   Up to 6 mm
○   Local cluster of corrosion
    at depths as above area
    up to 0.2 m²

1   Filling collector
2   Outlet collector
3   Outlet collector
4   Mixers

Figure 6.96   Corrosion damage in the tank in Figure 6.95 from the inside

Therefore corrosion on the bottom should be analysed as a process characterized by variable intensity, but occurring on both sides. This combined estimate of corrosion losses is of greater significance around the peripheral of the bottom zone.

Assessment of corrossive damage may be carried out when the tank is full, if acoustic emission methods are used. This involves registering by a set of sensors the sound waves generated by the leakage of liquid through the tank bottom or the corroded plates. However, using this method is difficult and so is not often chosen.

Recently developed methods of determining bottom damage caused by corrosion are based on measurement of steel plate thickness by ultrasonic techniques and the use of an appropriate measuring grid. With this method the tank has to be emptied. The method must be regarded as not very reliable, even if a dense measuring grid is used, because the point measurements do not guarantee that the areas indicated are the ones most exposed to perforation hazards.

Reliability in this respect is secured by measurements performed over all of the bottom. The measurements can be made by means of testers, which have been used for several years. These techniques can detect local plate thickness variations (corrosion pits), by the variable intensity in the magnetic flux generated by the tester. The tester is mounted on a trolley and looks like a lawn mower. Its relatively small size and weight of 50 kg enable it to fit easily through a side manhole

(Figure 6.97). Power is supplied either by a cable from a generator placed outside the tank or a battery inside the tester. The tester is equipped with a dozen sensors to monitor the magnetic flux variations caused by differences in the bottom plate thickness. Thus, by moving the tester over the bottom surface it is possible to register the plate thickness at every place on the bottom surface. Sensors can be freely mounted on the trolley. As a result of such mounting, the areas adjacent to the overlapping steel plate edges and other irregular surfaces can be inspected as well. Prior to measuring, the tester is calibrated by means of master templates whose depths correspond first to the total thickness of the steel plates used for the bottom at tank construction, and second to a thickness reduced by 40 – 50 per cent compared to the original plate thickness. The tester monitors thinned areas of sheet steel and detected losses are marked and then subjected to visual inspection and ultrasonic testing. The latest testers are equipped with a measuring system that makes it possible to localize and characterize failures, and to store them in the memory of a computer in a car outside the tank.

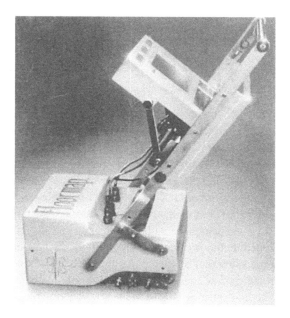

Figure 6.97 Tester for surface measurement of corrosion damage in the tank bottom

The most commonly used technique for measuring the sheet thickness of the tank shell, according to the most recent surveys is a series of ultrasonic measurements carried out at different places. Many measurements need to be made on particular shell sections to provide sufficient statistical data. This condition will be satisfied if a measuring grid of 1.0×2.0 m is used. On the basis of these results, the mean sheet thickness and standard deviation is calculated for each strake. The value accepted for strength analysis, i.e. the difference between the mean thickness and mean standard deviation, is a safe magnitude. The person who is to decide on the method and measuring technique should be able to access those shell areas that are particularly susceptible to corrosion. For instance, in tanks used for storage of crude oil, the area of most severe corrosion is usually the lower section of the plate strake at a height of about 300 mm above the bottom. Local corrosion pits in this area are deep and occur in lots (Table 6.1). Elsewhere corrosion usually occurs exclusively on the shell surface and its depth is insignificant [78].

The plate thickness of the shell can be measured by testers moving along the shell. These adhere to the shell by means of electromagnets, whose strength is enough to ride over local shell irregularities, such as oversize circumferential welds (Figure 6.98). The vertical movement of the tester is ensured by an electric motor

Figure 6.98 Tester creeping on the tank shell; 1 main view of the tank tested, 2 test device zoom, 3 control unit

driving a set of rollers. The distance covered by the tester is controlled from the ground by means of a joystick. The measurement results are shown on a digital display and in a form of computer printout. Plate thickness and measurement localization can thus be specified in detail. However, this method has not yet been widely used in Poland due to its high cost.

6.2.1.3 Determination of permissible settlement of the tank bottom circumference

The settlement of a tank depends on the type of foundation used. Until quite recently, tanks were usually set on a foundation of sand, encircled by a reinforced-concrete ring. The internal diameter of the ring was larger by ten centimetres than the diameter of the tank. Thus the load of the shell was transferred to a sand foundation of low elasticity. In the case of large capacity tanks with shells made of thick steel plates, this load is significant and can affect the settlement of the circumference of the bottom (Figure 6.99). This type of settlement is also characteristic of underground tanks constructed in a concrete lining where there is an additional loading from a layer of soil on the roof of the tank.

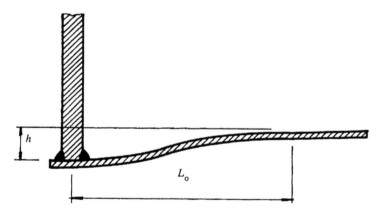

Figure 6.99 Typical settlement profile of the circumference of the bottom of a tank resting on a sand foundation

Figure 6.100 illustrates some values for the settlement profile of the bottom circumference of a tank of $V = 50000$ m^3 capacity, which needed repair after 20 years of operation. Settlement of the bottom circumference was responsible for additional bending stresses in this area, which in extreme cases can result in emergency cracking. Such an event was involved in a large leak that occurred in Edmonton in 1985, where a crack developed in the bottom of a tank of 12650 m^3 capacity [80].

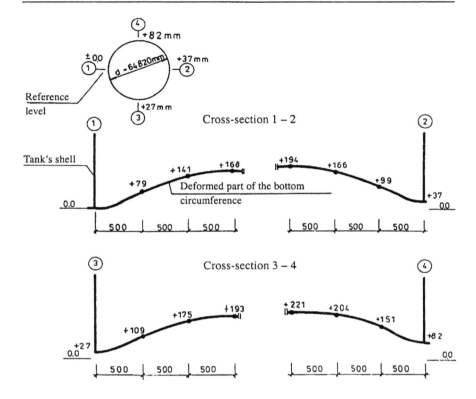

Figure 6.100   Bottom circumferential settlement of a tank of volume 50000 m³ after 20 years of operation

Circumferential settlement of the tank bottom can be detected by analysing its position in relation to the upper concrete edge of the surrounding foundation ring. However, the true settlement value and the profile of the deformed zone cannot be determined unless operation of the tank is shut down. After determining the stress values in the deformed zone, it is possible to decide whether the settlement poses any threat to the tank bottom. However, this decision requires significant and time-consuming calculation. The emergency condition of a particular tank with a crack in a deformed bottom can be roughly estimated using some graphs prepared by the Interprovincial Pipeline Enterprise [81]. The engineers of this Canadian firm, engaged to service 95 tanks of a total capacity reaching over 2 million m³, came to the conclusion that the procedure for determining the permissible settlement set out in API Standard 653 [82] is too conservative and worked out a method of their own. Taking advantage of a theoretical analysis and experimental investigations of the tanks in use, it was possible to plot a very useful graph that can help answer the question whether the tank settlement presents any threat to tank safety. Values that need to be taken into consideration are the radial lengths of the deformed bottom zone, at the place of maximum settlement. Two permissible settlement limits were assumed to be responsible for the deformation of the steel plates of the bottom: 0.3 per cent for non-corroded plates and 0.2 per cent for plates damaged by corrosion (Figure 6.101).

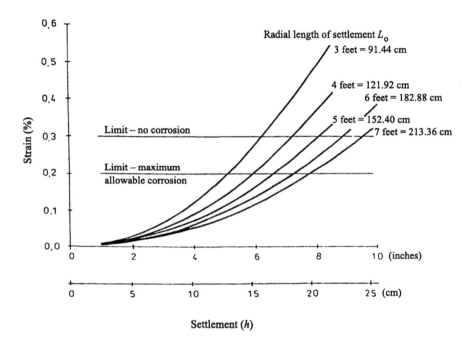

Figure 6.101   Graphs of the permissible circumferential settlement of tank bottom, according to [82]

The practical significance of this problem was pointed out in the article by J. Horner and R. Hinger [80] dealing with the repair of 13 tanks in Edmonton. The repair included replacement of the circumferential section of the foundation, the bottom plates and a low strake of shell plating. During the repair the tank shells were supported by hydraulic jacks. The procedure for repairing the foundations was changed with successive repairs; two of these solutions are shown in Figures 6.102 and 6.103.

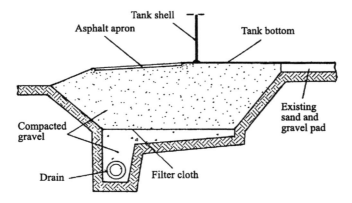

Figures 6.102 An example of repair of the foundation of a tank in Edmonton that had extensively settled under the bottom circumference (the first solution)

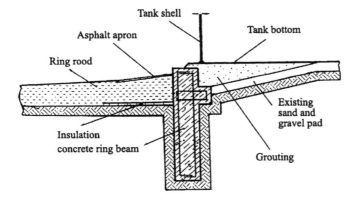

Figures 6.103 Other (later) solution for the foundation circumference repair of a tank in Edmonton

6.2.1.4 Methods of tank repair using welding techniques

Repair of the bottom

Assessment of what repairs need to be carried out on a tank can be made after a thorough removal of oil residues that have settled on the bottom and the lower sections of the shell. Any residue left on the bottom evaporates and produces dangerous concentrations of hydrocarbon vapours in the air, making it impossible to do any welding or cutting jobs inside the tank that involve open fire.

In a great number of tanks the bottoms are not flat, which is a result of using inappropriate welding techniques in constructing the tank. If the bottoms have bulges that have not been filled up with the sand, they are likely to deform when the quantity of fuel stored in the tank changes. Under such circumstances the plates in the bottom are subject to bending in the bulging zone. If, in addition, the plates are weakened by corrosion pits particularly continuous ones, the pitting can result in fatigue cracks. In the case of large bottom deformation, prior to painting or laminating the bottom with a plastic material it is necessary to carry out one of the following repair procedures:

- Cut through the bottom and stretch the plates using rigging screws to eliminate the bulging, then weld the cuts together
- Stabilize the deformed bottom by filling the bulge with bedding sand or concrete. The filling material is let into the bulging space through holes cut in the bottom. It is also necessary to make some air vents. Each of the openings is sealed with welded cover plates as soon as the bulging has been filled with bedding sand.

The first of these procedures was adopted in the repair of a tank of 32000 m³ capacity. Its diameter was 52.21 m and its steel plates had the thickness characteristics given in Table 6.1. In the bottom there were two large bulges (Figure 6.104).

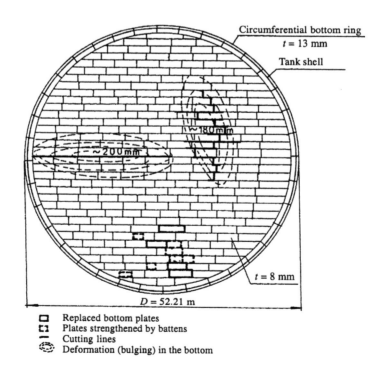

□	Replaced bottom plates
⊏⊐	Plates strengthened by battens
▬	Cutting lines
⊜	Deformation (bulging) in the bottom

Figure 6.104   Elimination of the deformation in the tank bottom and replacement of badly corroded bottom plates

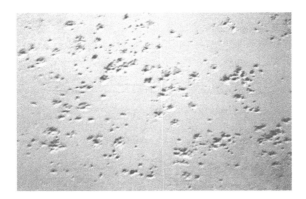

Figure 6.105   Cluster corrosion pits in bottom plates

The bulge located parallel to the strakes of the plates in the bottom was easy to eliminate. However, removing the bulge running diagonally to the alignment of the plates was more difficult and required a complicated cutting of the plates. In this tank, some of the plates were intensively damaged by clusters of point corrosion pits (Figure 6.105). The plates undoubtedly had some metallurgical defects, and there were non-metallic rolled inclusions on surface. It was decided to replace on entire plate if the pitting depths exceeded 5.0 mm, while if corrosion pits of 5.0 mm depth occurred only randomly, the damaged area would be repaired using cover plates. Four complete plates were replaced by new ones and seven others were repaired with cover plates (Figure 6.104).

Sometimes, on completing the sand blasting of the bottom of a tank, some greasy spots appear on the surface indicating leakage (Figure 6.106). In general, it will not be known when the perforation occurred, nor how much liquid has escaped to the sand foundation under the tank bottom. Welding repair works under such circumstances should be performed only when special safety measures are taken. During welding it is necessary to make it impossible for the fuel vapours that have saturated the sand foundation to escape through the perforation. In cases of point perforation, the holes should be enlarged with a drill to obtain circular perforations, which are then stopped by driving in pins of a lead bar. If the perforation is of an irregular shape, then prior to welding it is recommended that they be filled with a special type of paste used to seal leaking fuel tanks in military vehicles. In the welding area it is advisable to inject in some type of inert gas (e.g. CO_2 or nitrogen) into the space under the tank bottom. When the perforation is near the bottom perimeter, it is possible to drive a perforated pipe into the foundation mat to enable pressure injection of the inert gas. If the leakage occurs in central areas of the bottom, the operator should drill holes in the bottom plates to let in the protective gas. Cover patches welded to the bottom should have rounded edges so that the welded section does not make contact at a right angle.

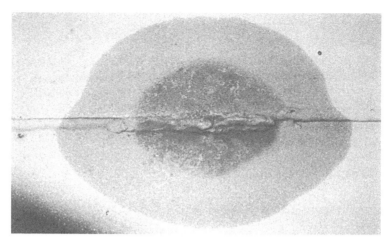

Figure 6.106   A crack in the weld of the tank bottom

Very often pitting corrosion occurs parallel to the welds linking the bottom plates (Figure 6.107). This is due not only to poor concentration of the weld, but also to changes in the structure and chemical composition of the steel in the heat-affected zone. The corrosion pits may range from 2.0 to 4.0 mm deep and need a furrowing and padding weld. Obviously all areas damaged mechanically in the course of the construction of the tank should also be padded.

A new bottom is hardly ever installed in a tank under repair. A procedure of that sort will be used only when pitting corrosion occurs over the entire surface

Figure 6.107   Corrosion pits along the weld in the bottom

area of the bottom, and the depth of the corrosion pits is greater than half the thickness of the plate used for constructing the tank. The decision to install a new bottom should be preceded by a thorough examination of the technical condition of the shell and roof to ensure that these structural elements have a useful life of least 20 years. Otherwise, in terms of technology and economics, replacement of the bottom is unjustified.

A new bottom will be constructed upon a sand mat, or a concrete layer laid on the old one. The thickness of the new layer is specified making provision for the new bottom to be fitted above the corrosion pitting in the shell (Figure 6.94). The design of the new bottom takes into consideration the possibility that the shape of the existing shell is not precisely a cylinder. Local deformations can reach several centimetres and it is difficult to identify them when the project is being implemented. For this reason, the layout adopted for the bottom plates should avoid the need to fit large plates inside the tank. The best solution is to design a circumferential ring for a bottom of width 250 – 300 mm that consists of plate sections not longer than 1500 mm (Figure 6.108). Small plates can be adapted to local irregularities in the shell shape by correcting their external edges on the building site. After the circumferential ring of the new bottom has been welded to the shell, the strakes of the plates in the main part of the bottom are connected, using a lap welding technique. The plates of the strakes can be joined together along their shorter edges on the outside of the tank, and when the qualitative weld test has been completed, these 'strips are' fitted into the tank through an assembly opening made in the shell (Figure 6.109).

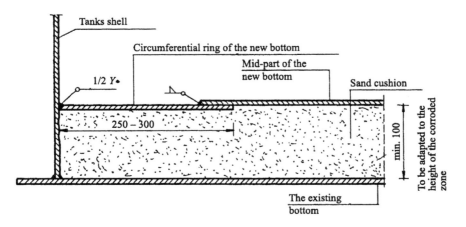

Figure 6.108 Additional bottom made during the tank repair

Figure 6.109 Assembly manhole in the tank shell

Another difficulty can arise in the construction of a new bottom in an underground tank. Most tanks of this type have a roof structure supported by a pipe pole located on the tank's axis. The pole is usually supported in articulated fashion by a cast steel bearing and the newly installed bottom must not be affected by changes in this support. Therefore a vertical ring made of pipe halves, cut along the pipes two generating lines is welded around the pole's circumference. In the next step, plates for the new bottom (Figure 6.110) are fixed to the ring by fillet welds. In an underground tank an assembly opening would be made in the roof and single plates for the new bottom let in one by one through this hole.

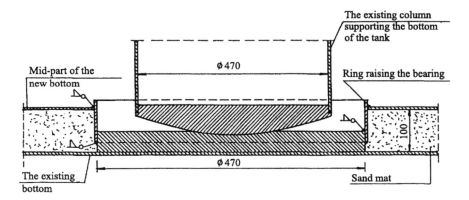

Figure 6.110   A detail of the joint in the new bottom joined to the bearing of the pole supporting the roof along the axis of the tank

New bottoms were installed in the riveted tanks. When the new bottoms are installed in tanks the tank equipment will often have to be mounted on higher position.

Repair of corrosion pits and delamination of plates in the shell

The largest corrosion pits in the shell occur in its lower parts, at a height up to 300 mm above the bottom (Figure 6.94). Such damage can be considered typical for tanks used for the storage of crude oil. A repair entailing strengthening of the shell is usually deemed necessary if the corrosion pitting has reduced the shell plate thickness to the minimum value specified in strength calculations.

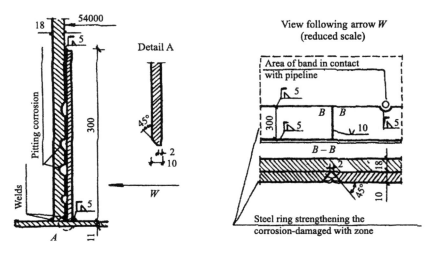

Figure 6.111   A band bracing the lower part of the shell damaged by corrosion

Under such circumstances a strengthening band is fitted either around the whole circumference of the tank, or along a section of the perimeter most damaged by corrosion most (Figure 6.111). The pits under the strengthening band do not need to be filled by pad welding. However, the band must be tightly closed by vertical and circumferential welds, because only if the oxygen supply to the corrosion pits is cut off will the corrosion process be brought to a halt. In cases where the fillet weld connecting the shell to the bottom is thick enough, it is possible to make the strengthening band out of two plates (Figure 6.112). In such a solution the band will not be too thick.

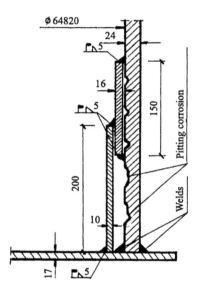

Figure 6.112   Double-plate band bracing of the shell – the solution recommended when the fillet weld joining the shell to the bottom is large

In the upper sections of the shell, is sometimes noticed plate delamination when plates are blast cleaned. Delaminations are dangerous with respect to strength loss if they run diagonally to the plate surface. Such delamination puts part of the plate out of operation, and consequently the damaged area has to be cut out of the plate. The extent of the delamination is determined by ultrasonic tests carried out at a high density of measuring points (Figure 6.113). The cover plate prepared for welding in place of the delaminated fragment cut out should have slightly larger curvature than that of the shell. This is necessary to compensate for welding shrinkage following the execution of the vertical welds. These vertical welds are always carried out first because of their significance for primary strength. Transverse shrinkage of vertical welds will result in a partial straightening of the welded plate, so to avoid local post-welding shell deformation, the cover plate needs the appropriately larger curvature before welding work slabs. The larger curvature of the cover plate can be calculated using data from a welding handbook for transverse shrinkage of a butt joint of a given thickness, using appropriate welding techniques. In one of the tanks under repair, it was necessary to cut a section 700 mm wide out of a 11 mm plate strake. According to the calculations, the curvature radius of the 'patch' plate had to be 3 mm larger than the shell's original curvature radius (Figure 6.114). On completing the welding work, it turned out that the patch fitted the tank shell perfectly.

(a)   (b)

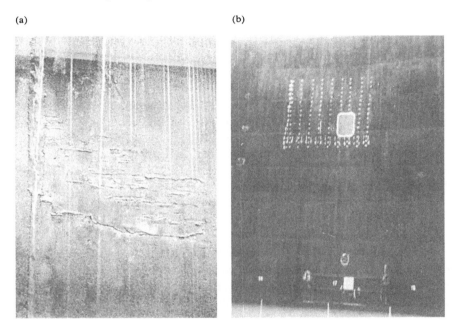

Figure 6.113   Delamination of plate in the tank shell: (a) view after sand blasting, (b) extent of delamination detected by ultrasonic testing

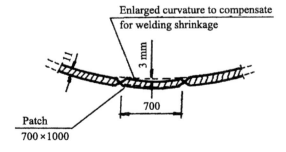

Figure 6.114   Initial bending of a plate welded to the tank shell to replace the delaminated fragment cut out

Correction of the shape of the shell of a tank

Correction of the shape of the slab of a cylindrical steel tank is usually needed for three reasons:

- Mistakes made during shell assembly and welding
- Mistakes made during the operations (such as negative pressure generated in the tank [83]);
- Non-uniform settlement of the tank foundation pad.

These events often occur together and may cause one another:

- Local deformations of the shell shape caused during assembly reduce the value of the critical negative pressure that is responsible for the loss of the shell stability and the operational capability of the tank
- Non-uniform settlement of the foundations while the tank is being constructed results in an accumulation of irregularities in the shape of the successively assembled strakes of the shell plates.

The elimination of the shell deformation is indispensable for tanks with a floating roof. Deformations involving convexity towards the interior of the tank may hamper free vertical movement of the floating roof. In tanks with permanent roofs, shell deformations must be eliminated if the plates have sharp edges or cracks. However, improving the shape of a cylindrical steel tank shell is a technically difficult task, for at least two reasons:

- Local deformations may be redistributed after cutting through the shell
- Deformations may occur as a result of the inaccurate fitting of substitute plates into damaged areas.

The state of stress in a locally deformed cylindrical shell is not easy to determine. However, the stresses causing the deformation will remain in a state of equilibrium as long as the integrity of the shell is not disturbed. When damaged or excessively deformed plates in the cylindrical shell are cut, the integrity of the shell is broken and the area weakened. This results in a displacement of the existing deformations and occasionally also the occurrence of local strains in parts of the shell whose shape was previously correct. The redistribution of the local deformation can lead to the development of such large bulges over or under the cut-out opening in the shell that welding the plate to close the opening will require the use of very complicated hot overbending methods. To avoid this, the tank shell needs to be braced before cutting, using double-tee bars fixed above and below the opening to be cut out.

Another technically complicated operation is fitting the plate or plates to replace the cut-out fragment by butt joint welds to the shell. The shape of these plates must fit the existing hole perfectly. If just one of the welds has a variable thickness along the contact length, the tank will locally deform again. The best solution is to prepare a marking-off template, matched to the shape of the cut-out hole in the shell. This can be used to mark out the plate that will fill the hole in the tank shell. When specifying the dimensions of this plate, the welding shrinkage that will result from vertical welds must be taken into consideration. Using a marking-off template will work, if it is necessary to cut out a large fragment. It is better to make use of lap joining technique when damaged or badly deformed plates are cut out of the shell. However, the use of lap joints is not possible in tanks with floating roofs or when the cut-out plate is several millimetres thick.

If the tank shell has been deformed by negative pressure generated in its interior, an initial repair of the shape can be done by means of hydraulic expansion. This is not a new solution; such a method has already been discussed in several publications, e.g. [83], [84], [85], but the results of expansion operations have only been judged subjectively – no measurements were carried out before and after, such as by filling the tanks with water.

The case described below is a notable example due to the fact that despite the shortage of time (the tank was to be put into operation as quickly as possible), the survey of the shell shape was carried out for three stages:

- After an emergency deformation
- After filling the tank with water
- After the tank had been emptied.

Comparison of the three surveys gives an idea of the effectiveness of the hydraulic expansion technique when used in an emergency repair to a shell damaged by negative pressure generated in the tank.

Figure 6.115   Shell deformation of a tank of volume 196 m³

Figure 6.116   Shell deformation of a tank of volume 196 m³

The recommendations given above can be illustrated by four examples of repairs to tanks shells deformed in various ways. The first of these analyses the hydraulic expansion of two deformed tanks.

Example 1 Two identical cylindrical and vertical tanks with permanent roofs were damaged. The tanks were newly built for a refinery complex [86], with a capacity of 196 m³. Their main dimensions were: shell internal diameter $D_w = 5.00$ mm and shell height $H = 10.00$ m. The tank shells consist of six strakes of steel plate of 7, 7, 6, 6, 6 and 5 mm (from the bottom). The tank bottom is flat, while the conical roof structure is made up of eight radially arranged ribs of 120 double T bar. On the tank axis the ribs are joined with a jointer, and additionally at every one third of their length they are connected by purlins, creating two polygonal rings. The tanks were made of regular quality steel whose yield point is 235 MPa. The shells were deformed when water was rapidly pumped out during a hydraulic test to determine the strength of the newly built tanks. Air was not supplied to the emptied tanks, and as a result on the both shells some alternating concavities and bulges appeared. The largest concavity was approximately 300 mm deep, while the arch of the largest bulge was 700 mm. No cracks occurred in either the tanks shells or plates. However, some deformations were accompanied by sharp-edged plate bumps. The shell deformations are shown in Figures 6.115 and 6.116.

Shell deviation from plumb-line (mm)

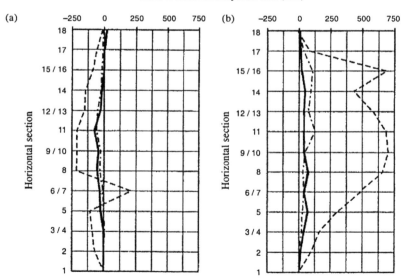

Figure 6.117   Comparison of deformations in the tank shell along the measuring plumb lines: (a) No.4, (b) No.5

- - - - - - - - -   condition after damage

_._._._._._.   condition after filling the tank with water

_____   condition after emptying the tank (on completing the hydraulic expansion procedure)

Ten plumb-lines measures were marked on one of the tank shells, but on the other only nine, because of existing installations. Each plumb-line was provided with 18 measuring levels, three on every strake of plates. It was possible to carry out deformation surveys three times, for each of the three stages distinguished above. The resulting measurements, which demonstrate the efficiency of the hydraulic expansion technique, are given for two selected measuring plumb-lines (Figure 6.117). The results for tank after it had been cleared of water and subjected to hydraulic expansion were used to draw a map of the deformations.

Apart from the shell's generating lines on which the surveys were made, the deformation values for the remaining surface were determined making use of computer interpolation (Figure 6.118). This map was used for assessing what repairs were necessary. It was decided that shell fragments that were deformed more than ± 30 mm were to be cut out. Figure 6.119 shows this part of the shell. Vertical cuts were made at a minimum distance of 300 mm from the existing vertical welds. New shell patches were fitted successively in respective strakes, running from the bottom to the top sections. The plates to be inserted in the holes were given by a larger curvature than that of the shell. This was done to avoid welding shrinkage in the vertical welds and so to obtain the correct cylindrical shape after the new plates had been welded in, without local deformations of the shell.

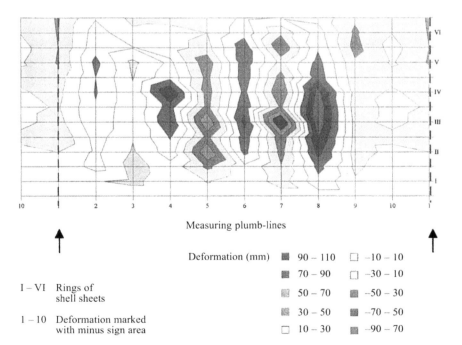

Figure 6.118 **Map of deformation in the shell after hydraulic expansion**

The repair carried out using this method produced good results. The process of hydraulic expansion significantly reduced the surface area of plates that need to be replaced in the repair.

Example 2 This involved a tank shell with a fixed roof [87]. and capacity of 2760 m³. The vertical cylindrical tank has a flat bottom and a dome-shaped roof. The main dimensions of the tank are: shell diameter D = 18.30 m; shell height H = 10.50 m; and roof dome rise arch f = 1.30 m.

The tank is of welded construction and is made up of seven strakes of 10, 8, 7, 6, 5, 5 and 5 mm thick, respectively. The upper edge of the shell is shaped as a firm coping. Inside the tank a bar 100 × 10 has been welded to the shell and there is a channel section 120 loaded on its external side. The roof load-bearing structure consists of 20 arch ribs (I 160) of span approximately equal to the shell radius. In the tank centre the ribs are connected by a coupling ring, and there are five purlins along the ribs, located at 1.45 m intervals. Every fifth rib in the roof structure has a truss bracing; altogether there are four truss braces. The roof cover is made of 3 mm thick plate. Due to an error during operation, negative pressure was generated in the tank. As a result, the shell was deformed at two places around the tank circum-

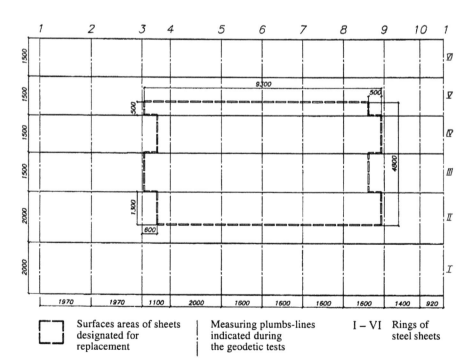

Surfaces areas of sheets designated for replacement | Measuring plumbs-lines indicated during the geodetic tests   I – VI   Rings of steel sheets

Figure 6.119 Zone of deformed shell requiring replacement, prepared on the basis of the deformation map

ference. Some more extensive deformations occurred below the service platform connecting the roof of the tank to another nearby tank. Deformations were found on the second and third plating strake from the top (both plating strakes were 5 mm thick), and extended over a length of approximately 12 m (Figure 6.120). The arch rise of the concavities and convexities amounted to about 150 m, and in the second plating strake from the top, there was a crack along a sharply bent plate of length approximately 140 mm (Figure 6.121). Deformations in the second strake of the shell circumference were sig-nificantly smaller and of a less destructive nature. The roof was not deformed owing to its rigid construction (described below). However, the tank, though was generally in good technical condition, required repair. The deformed plates were to be cut out and the new plates fitted. Hydraulic expansion techniques could not be applied because the deformations had sharp edges and there was a crack in one of the plates. From the repair assessment it was decided that the plates 12 m long in two strakes of the shell (i.e. those at 3.0 m height) should be replaced.

Figure 6.120 Tank shell deformations caused by negative pressure

Above the cut-out hole there were five ribs in the roof-bearing structure, so three of them that were situated in the centre had to be supported by a scaffold prior to the commencement of the disassembly work on the damaged plates (Figure 6.122). The tank shell was braced by a double T bar 300 mm below the opening to be cut out.

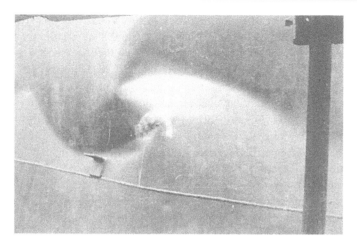

Figure 6.121   Shell plate cracked along its sharp edge

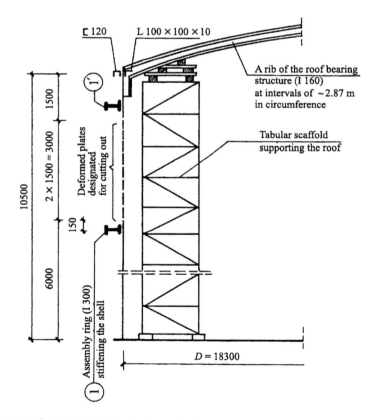

Figure 6.122   Protecting the tank shell and roof during the repair

The upper edge of the opening was about 1500 mm below the solid ring coping of the domed roof cover. Thus, the tank shell in this zone was rather rigid. Therefore, the assembly stiffening ring made of a double T bar was used only in the area near the vertical edge of the hole to be cut out (Figure 6.123). In the neighbouring area of horizontal contact between the plates welded in the second and third plating strake of the shell, an auxiliary assembly bracing ring was fitted. The plates in the third strake from the top were butt welded. The plates closing the cut-out opening, i.e. the plates in the second strake from the top, were joined by means of overlapping welds to simplify fitting (Figure 6.123). The repair of the tank was carried out successfully and from the technical point of view, the overall result was positive.

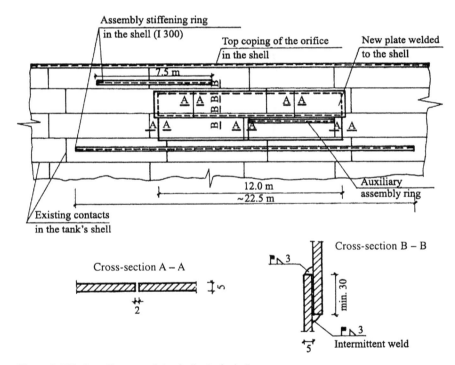

Figure 6.123   Inserting new plates in the tank shell

Example 3   This involved the repair of the shell of a tank with a floating roof. The tank is of 50000 m³ capacity and its other main dimensions are internal shell diameter $D = 65.0$ m and shell height $H = 18.0$ m.

The tank shell is made up of nine strakes of plates of thickness 25, 22, 19, 17, 16, 16, 12 and 9 mm respectively from the bottom. The floating roof is of a pontoon-membrane type and has a mechanical seal.

The tank was made in 1975 and had been in operation for 15 years, though only 70 per cent its capacity was in use due to deformation in a form of a fold with an inward bulge in the upper circumferential part of the shell. Surveys indicated that the rise of the arch in this deformation at some points reached 370 mm, and was in the range of 100 – 200 mm over nearly the entire circumference of the shell (Figure 6.124). The cause of such an unusual deformation could not be established at the time of repair preparation, but repair was still necessary because the floating roof could not move above the deformation area.

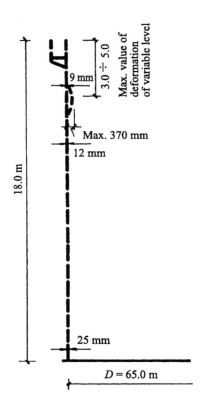

Figure 6.124   Shell deformation of a tank of volume 50000 m³

Before the repair commenced, it was decided that the cheapest method should be used: cut out the deformed plates and replace them with the new, properly shaped ones. The extent of deformation indicated that the repair might be difficult. The owner of the tank decided to have the damaged plates replaced by new ones. As the tank had a floating roof and the most appropriate plates for replacement were 9 and 12 mm thick, butt joints had to be used. However, precise fitting of the plates to a cut out hole of given dimensions appeared to be very difficult. The welds did not have the same thickness along their entire length. This resulted in shell deformation, in spite of attempts and different welding methods used to avoid it: heating and cutting in the plate bulges were tried. After three plates had been replaced, this method was abandoned as unsatisfactory.

An alternative engineering method to repair such a tank is shown in Figure 6.125. The more deformed peripheral half of the shell was reinforced by a flange weld using a double T bar ring at a height of 3.5 m below the service walkway.

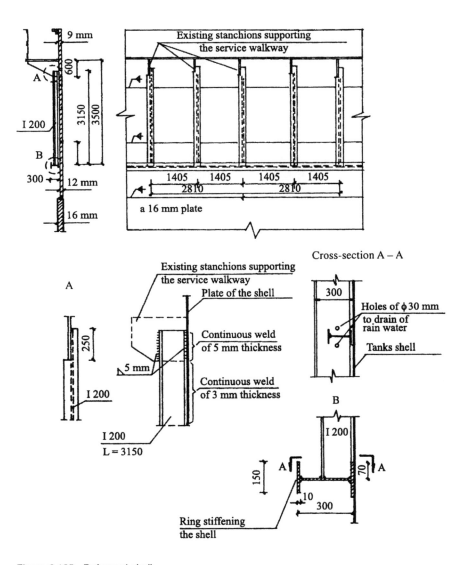

Figure 6.125   Deformed shell

This ring was connected to the service walkway by means of stanchions made of a double T bar 200. The stanchions were spaced around the shell at intervals of 1405 mm. The stanchions were applied to tighten up the deformed plates of the shell by a yoke with set screws (Figure 6.126). In the area that presented more difficulty in eliminating the deformation, the spacing of the stanchions was closer (Figure 6.127). When the shell plates were fully stretched to the stanchions, continuous welds were made to join the plate to the foot of the double T bar used for the stanchion. The welding was performed using a step-back method.

Figure 6.126   Pulling the shell plates tight to the bracing stanchions

Figure 6.127   View of the tank in the final stage of repair

The use of the intermittent welds, although fulfilling the strength requirements, proved unsatisfactory due to conditions favouring rapid development of corrosion in cracks, which were difficult to avoid in such circumstances. To inhibit corrosive processes, several drain holes for rain water were drilled in the web of the I-beam shell profiling ring. Although the repair work was limited to half the shell perimeter only, it was possible to obtain undisturbed movement of the floating roof over the full height of the tank shell.

Example 4 The floating-roof tank to be repaired is of 10000 m³ capacity [12], with main dimensions internal shell diameter $D = 35.85$ m and shell height $H = 16.0$ m.

The tank shell is made up of eight plating strakes of plates of thickness 16, 14, 12, 11, 9, 8, 7 and 7 mm. The shell is coped with an $80 \times 80 \times 8$ mm double T bar. The peripheral flange to reinforce the shell, used also as a service walkway for the maintenance staff, is welded at a distance of 1.1 mm to the upper edge of the tank. The tank's bottom thickness at the periphery is 10 mm and in the centre 9 mm.

The floating roof is of a pontoon-membrane type. The tank rests on a sand foundation, rimmed with a reinforced-concrete flange 25 mm wide. The internal diameter of the flange is larger by 180 mm than the diameter of the tank. The tank was built on the ground frequented by mining disasters. The foundation sand under the tank may have been improperly consolidated, for the tank settled unevenly and the upper part of the shell became too oval-shaped. During the water acceptance test, the floating roof stuck at the height of 10 m above the bottom (i.e. 6 m below the top of the shell).

(a)

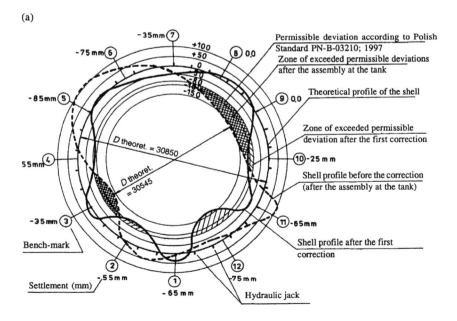

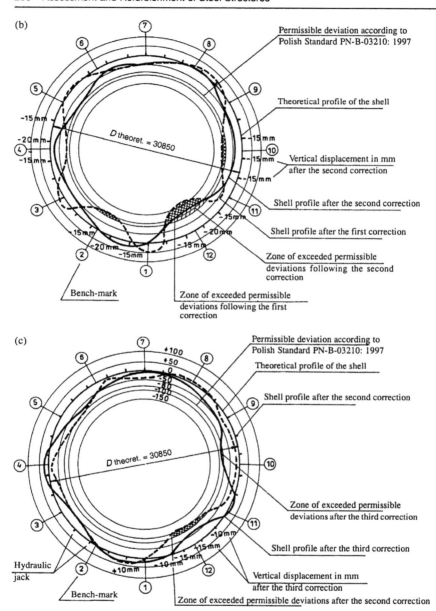

Figure 6.128 Shape of the upper edge of the shell in a tank of volume 10000 m³ (deformations have been scaled up in relation to the tank diameter) (a) condition after the assembly of the tank and after the first correction; (b) condition after the first and second correction (the jacks with no displacement date were not put into operation in the correction procedure; (c) correction after the second and third correction (the jacks with no displacement data were not put into operation in the correct procedure)

The results of the survey carried out after the damage made it possible to prepare a map of the shell shape at its respective levels. The shape of the upper edge
is indicated by the broken line in Figure 6.128(a). The figure also illustrates the
settlement values for the tank, measured from 12 datum points equally spaced
around the perimeter of the bottom. Comparison of the upper edge of the shell shape
with the settlement of the tank gives rise to the following remarks for further consideration:

- The largest settlement is at points located at the end of two diameters, situated closely to bench-marks 5 and 11 (settlement of 85 mm and 65 mm), as well
  as bench-marks 6 and 12 (settlement of 75 mm and 75 mm)
- The largest elliptical elongation of the upper part of the tank shell is along
  the same diameter
- An improvement in the shell shape can be achieved by value differentiation,
  the vertical displacement of respective parts of the tank's perimeter, a meth
  od that has been applied successfully elsewhere.

Correction of a tank's shell shape by adjusting the level of the foundations can
be very effective, provided that strain freedom in the upper part of the shell is
ensured. To fulfil this condition in the repair of this tank it was necessary to cut off
both the service walkway in the top strake of the shell for the maintenance personnel, and the $80 \times 80 \times 8$ mm double T bar coping of the shell. A cylindrical tank
shell with no peripheral stiffening of the upper edge can be deformed by wind

Figure 6.129   Tank shell secured by guys against wind load deformation volume

action [84], [89]. For this reason, prior to disassembling the maintenance personnel service walkway, the shell was braced by 12 guys. They were attached to the stanchions of the service walkway later to be cut in pieces, and to an anchor fixed to the ground (Figure 6.129). The shell was lifted by 30 hydraulic jacks making use of specially designed fixtures. The jacks were set on the peripheral concrete ring rimming the foundations. Fixtures for lifting the tank should not be welded to the tank shell. Any form of indentation in the lowest plating strake of the shell, which is particularly vulnerable to brittle cracking, must be avoided.

Welding of the assembly elements in the shell causes structural notches and may lead to the initiation of substantial notches during the final repair stage, when the assembly elements are being cut off, and to burnt areas in the shell plating. Therefore the best solution for lifting the shell would be to use a fixture with a horizontal bearer, which can be inserted under the bottom (Figure 6.130). The advantage of this has already been demonstrated elsewhere. Unfortunately, it was impossible to make use of this solution with the tank under discussion, due to an insufficient distance between the bottom perimeter and the internal edge of the ring enclosing the foundations. Thus the fixture illustrated in Figures 6.131

(a)   (b)

Figure 6.130 Assembly fixture applied to lifting the shell of another tank (volume 20000 m³) after uneven settlement was detected: (a) main view; (b) construction of the fixture with a horizontal bearer inserted under the tank bottom

and 6.132 was used. Its lower part was welded to the bottom circumference by means of two plates (1), while its upper part was supported on the shell using a roll-shaped plate (2). The fixture was enclosed in a steel frame and suspended from it by two tension members (3). In this way the horizontal position of the perimeter to the bottom was stabilized at the proper level. It is important during surveys taking several days, and also after they have been finished, that while the cavity between the tank bottom and the surface of the foundations is filled with gunite, the circumference of the bottom remains at a constant level, in case the hydraulic jacks break down.

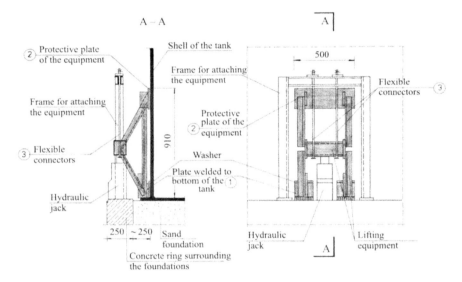

Figure 6.131   The fixture used for lifting the tank shell

Figure 6.132   View of the fixture in Figure 1.131 (an indicator for controlling the lifting height; the indicating dial has already been deflected)

Adjusting the level of the shell foundations was performed in three stages. The first was to eliminate the deformations that were the result of uneven settlement of the tank. The shell was therefore lifted so that the relevant bench-marks could be displaced by the value of the settlement established by them, i.e. plus 80 mm, which correspond to the diameter of the flexible hose to be used later to fill the empty space between the tank bottom and the foundations with gunite. However, the theoretical vertical displacements calculated for the respective points on the shell perimeter could not be properly achieved because of significant rigidity in the shell, and the vertical displacement variations forced by the jacks near them could not be made greater in practice than 5 to 10 mm.

Both the theoretical and actual programmes of vertical displacements for the various jacks are shown in Figure 6.133. On completing the first stage in correcting the settling of the tank, the shape of the upper line of the shell was highly unsatisfactory (Figure 6.128(a).

(a)

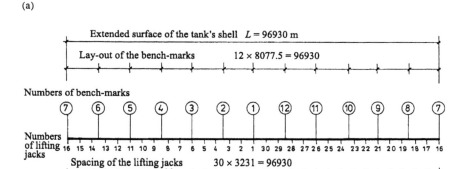

(b)

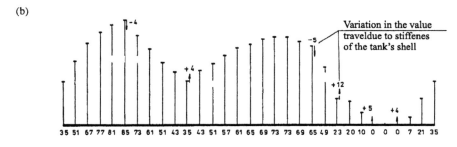

Values of theoretically necessary travel of the jacks pistons

Figure 6.133 Theoretical and practical jack displacement values in the first correction test: (a) arrangement of bench-marks and jacks; (b) piston displacement values of the hydraulic jacks; theoretical and real values

In the next two stages in adjusting the foundation, only selected hydraulic jacks were used (Figure 6.128(b) or 6.128(c)), and the shell was brought into a shape that enabled the floating roof to move freely over the entire height of the tank. Although the shape of the upper edge of the shell still deviated from the theoretical circular shape, it is very important that the permitted negative deviation (towards the tank interior) is exceeded by only 10 mm along an insignificantly short length of the perimeter in the area of bench-mark 10. However, this did not endanger the smooth movement of the roof, which had been proved by the second water test of the tank.

The difficulty in obtaining the correct circular shape of the upper edge of the tank is related to significant deformations along the various generating lines of the cylindrical shell. Two characteristic cross-sections of the shell illustrate this: the cross-section between bench-marks 3 to 9 after the assembly of the tank had the smallest diameter (Figure 6.134(a)). Cross-section 5 to 11 (Figure 6.134(b)) is close to the largest deviation of the upper shell edge after assembly of the tank. Apart from the substantial irregularities in the shape of the shell along the generating lines, Figure 6.134 also shows the effectiveness of the repair method used. The range of deformation for different heights of the shell at the various repair stages is shown for sections 3 to 9 and 11 (Table 6.2).

(a)   (b)

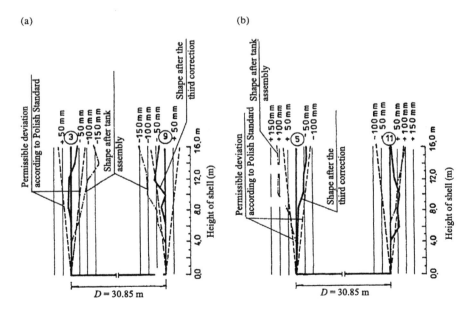

Figure 6.134   Deformations along the characteristic generating lines of the shell after completing the tank assembly and after the third correction test: (a) a cross-section at bench-marks 3 and 9 (b) a cross-section at bench-marks 5 and 11

Table 6.2 Out-of-plumb values for shell generating lines, after assembly of the tank and successive corrections of the tank shape

Measuring (above bottom) m	Value of permissible deviation cm	Out-of-plumb values measured at different repair stages* cm								Difference between the theoretical diameter and the real one, between the bench-marks 3 and 9.5 and 11 cm	
		Generator 3				Generator 9					
		A	B	C	D	A	B	C	D	A	D
16	8.0	−14.0	+2.5	−4.0	−4.0	−16.5	−3.0	−4.5	−6,5	−30.5	−10.5
14	7.0	−17.0	+4.0	−2.5	−7.0	−13.0	−4.0	−5.0	−6.0	−30.0	−13.0
12	6.0	−11.0	+5.0	+0.5	−3.5	−12.0	−2.0	−6.5	−5.0	−23.0	−8.5
10	5.0	−9.0	+5.5	+1.5	−2.5	−11.0	0.0	0.0	−2.0	−20.0	−4.5
8.0	4.0	−6.5	+4.5	+1.5	−1.5	−6.0	−1.5	−2.0	−3.0	−12.5	−4.5
6	3.0	−4.5	+1.5	0.0	−2.0	−4.0	0.0	0.0	−0.5	−8.5	−2.5
4	2.0	0.0	0.0	0.0	0.0	0.0	0.0	0.0	0.0	0.0	0.0
		Generator 5				Generator 11					
16	8.0	+12.5	−1.0	−6.5	−5.0	+9.0	−4.5	+1.0	+1.5	+21.5	−3.5
14	7.0	+11.5	−2.0	−6.5	−4.0	+7.5	−2.0	+2.5	+3.0	+19.0	−1.0
12	6.0	+11.0	−1.0	−4.5	−1.5	+5.5	+0.5	+3.5	+6.0	+16.5	+4.5
10	5.0	+10.5	−1.5	−4.0	−2.0	+4.5	+4.0	+6.0	+6.0	+10.5	+4.0
8	4.0	+5.5	+1.0	−1.0	+0.5	+3.5	+4.0	+5.0	+5.0	+9.0	+1.0
6	3.0	+3.0	+1.5	0.0	+1.0	−0.5	+5.0	+5.5	+5.0	+2.5	+6.0
4	2.0	0.0	0.0	0.0	0.0	0.0	0.0	0.0	0.0	0.0	0.0

* A   After assembly of the tank
  B   After the first correction, on completing the lifting of the shell by a value corresponding to the settlement noted at respective datum points (compare Figure 6.9(a))
  C   After the third settlement correction of the shell
  D   After filling the cavity between the bottom of the tank and the foundations (when the wind ring was welded and the next water test finished).

Following the third stage in correcting the tank shape, when the result obtained was judged to be satisfactory and an assembly ring made of a double T bar 220 was installed inside the tank to stabilize the top edge of the tank shape, the service walkway around the highest plating stake was re-welded to the shell.

After the floating roof seized up during the first water test, the water was left in the tank for over six months. The reason for this was to consolidate the sand in the foundations and to end its settling. As soon as the shape of the tank had been corrected, the cavity between the bottom and the foundations surface was filled with gunite at a pressure of 0.5 MPa. Approximately 17 m³ of gunite were used for this purpose.

Adjusting the level of the tank foundations had caused a change in the shape of the periphery of the bottom that made it necessary to cut out a section of approximately 1.0 – 1.5 m along the radius. As a consequence, the supports of the floating roof needed some mechanical re-working: their length had to be adapted to the new profile so the roof would not bend when it was resting on the supports.

Repair of a fixed roof damaged by negative pressure generated in the tank

Negative pressure is often generated in a tank when the product stored in it is pumped out of the tank and the outflow of the fluid in not counterbalanced by an equal volume of inflowing air. This is likely to occur if the emergency valves installed in the tank roof malfunction or if their operating efficiency is insufficient. In the case of a roof with a large surface area, even if the negative pressure is relatively insignificant, it can cause a load to be destructive to the roof surface or to the top part of the shell. The element that is less rigid or has some geometrical imperfections that initiate local loss of stability will be destroyed.

As a result of the negative pressure produced in the tank, damage is inflicted on the roof (Figure 6.135(b)), or the top part of the shell (Figure 6.136). If damage occurs to the roof, then any destruction initiated locally, will usually spread over the entire surface of the roof, causing it to fall in. Due to the structural phenomenon of snap through the roof shape from convex (Figure 6.135(a)) to concave

(a)

(Figure 6.135(b)). If the roof destruction process is noticed early enough and an adequate quantity of air is let into the interior of the tank, the scope of destruction can be limited to a part of the roof, as in the case of a $V = 5000$ m³ tank (Figure 6.137).

(b)

Figure 6.135   Tank roof of volume 5000 m³ (a) before damage, (b) after damage

Figure 6.136   Upper part of the shell of a tank 5000 m³ damaged by negative pressure produced in the tank

Figure 6.137   Tank roof of a tank of volume 5000 m³ damaged by negative pressure

However, a damaged roof does not lose its bearing capacity. If the roof has completely fallen in, the structure continues to operate like a suspension structure and the sunken roof can be useful as a working platform from which to rebuild the load-bearing structure of the roof. If only part of the roof has fallen in, it is possible to prop it at the centre of the tank by a tabular supporter, and also use the destroyed roof as a working platform in the reconstruction work. The upper shell edge of tanks with dome-shaped roofs is stiffened by a coping ring, usually of large cross-section. The ring protects the shell from transferring the deformation from the roof to the shell. If the shell has not been affected by deformation, it is recommended that a superstructure be added to the shell for example in form of a ring of plate ranging from 500 to 750 mm in height. The ring on the new top edge of this shell is designed as a coping the new roof structure (Figure 6.138).

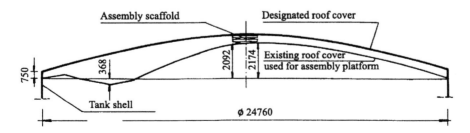

Figure 6.138   Repair method for a damaged tank roof cover, tank of volume 5000 m³

A schema and a general view of the repair work in progress are given in Figure 6.139. The ribs of the new roof structure were moved by half the spacing of the exiting ribs to facilitate their disassembly after completion of the new roof structure. In the new roof-bearing structure the purlins were not mounted close to the load-bearing arch ribs of the roof. The intention was to leave an opening for the segment of the old roof, which were left temporally on the tank bottom, to be removed later through the old roof when the repair work was completed (Figure 6.140). The assembly pole used to support the connecting elements of the damaged old roof was also pulled out through the same hole.

(a)

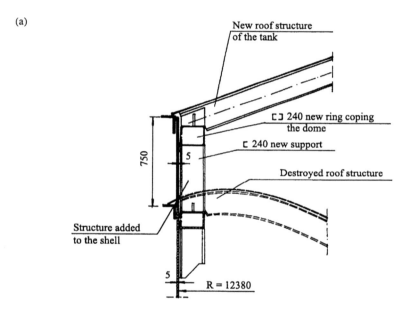

(b)

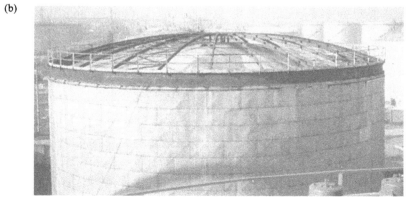

Figure 6.139 Repair of the roof of a tank of volume 5000 m³: (a) schematic, (b) the new roof-bearing structure assembly, final phase

Figure 6.140   Travelling crane waiting to remove the cut out elements of
the old disassembled roof

6.2.1.5 Non-welded repairs of tank bottoms and shells

Corrosion damage to the bottom of a pitting – may be well protected against further corrosion by a laminate coating made of synthetic resins, reinforced with an
artificial fibre mat or fabric. When the depth of the corrosion pits is less than about
20 per cent of the thickness of the middle bottom plate, padding of the pits before
laying the laminate is not required. The depth of corrosion pits in the shell that are
not to be pad welded has to be calculated on the basis of an analysis that indicates
both the corrosion pit depth and their cross-section. The laminate coating effectively prevents contact between the surface of the plate and water beneath the
product, and consequently halts the corrosion process.

Currently available laminates are made from two-component paints (solvent-
less) and artificial fabrics or mats suitable for a given paint. Laminates laid on
properly cleaned and roughened surfaces (cleaned by sand blasting to degree of
purity SA 2.5) adhere to steel surfaces very well and retain durable plasticity. This
is of major importance, because many tank bottoms are not perfectly flat. Local
deformations of the bottom are displaced (by alteration from convexity to concav-

ity) with change in the pressure of the liquid column on the bottom during empty-ing or filling of the tank. Laminate paints also have to be good conductors of static electricity generated by operating the tank structure. Graphite added to the paint usually helps to achieve this conductivity.

The process of laying a laminate on the bottom and shell of a tank is divided into several phases:
1 Sand blasting of the steel surfaces to a grade of purity that mats SA 2.5. The abrasive material used must have the proper granulation and chemical purity so as to produce a steel surface of adequate roughness.
2 Abrasive material is removed after sand blasting by means of a heavy-duty vacu-um cleaner
3 Grinding of sharp edges and remnants of mounting handles and the welding fitt-ings. Paint adhesion would be unsatisfactory on these surfaces and this could give rise to cracks in the protective layer.
4 Very careful removal of dust from the tank.
5 An examination of the tank to find places with deep and extensive corrosion pits, microperforation and macroperforations.
6 Repair of perforations by filling with lead stud.
7 Manual painting of complicated roof structures, pipes, valves and weld joints in order to cover the surface very thoroughly before airless spraying, which could prove too thin if the initial painting was not done manually.
8 Inspection of the repaired perforations to detect leakage.
9 Laminating of deep perforated corrosion pits, or the whole of the bottom with a glass mat of density $200g/m^2$. If the whole bottom is to be laminated, the lami-nate is laid over the shell up to a height 700 – 800 mm, i.e. above the below-product water layer.
10 Application of protective material by means of hydrodynamic spraying to achie-ve a dried layer thickness of:
   • 300 μm – for a new tank
   • 500 μm – for an extensively used tank.

When executing all procedures connected with laminating and after the lamina-tion hardening period, it is necessary to maintain constant temperature and humid-ity by means of air conditioning. This is a vital element of the engineering process, which influence the quality and durability of the achieved surface.

Professional laminating companies usually give a ten-year guarantee for lami-nation jobs. This process technology has been improved by using multi-layer lin-ing, shown in Figure 6.141, and using an overpressure monitoring system on the 3D fabric. The 3D fabric is composed of two parallel structural layers, joined by densely packed vertical fibres. The gap (of width from 3 to 24 mm, depending on the fabric) created by means of capillary forces between the structural layers when the resin saturated, is used to monitor the leakage.

Figure 6.141   Multi-layer anti-corrosion coating with leakage monitoring:
   1 the tank bottom,
   2 lower layer of 3D fabric stuck to the tank bottom with the epoxy resins,
   3 space between the upper and lower layer to monitor leakage,
   4 upper layer of 3D fabric,
   5 antistatic layer resistant to the stored medium

The 3D fabrics most frequently used to protect fuel tanks, have gaps of 3 to 4.5 mm. When the resin has been hardened sufficiently, the 3D fabric may be loaded to a high pressure, considerable exceeding the pressure exerted by the stored product, augmented by the control pressure. If the upper part of the fabric or the bottom plate in the monitored space has been perforated, the pressure changes and the outside sensors will respond by emitting visual and acoustic signals.

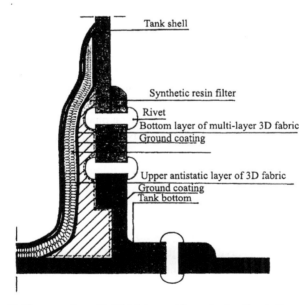

Figure 6.142   Multilayer coating with 3D fabric used for protecting the shell connection with the tank bottom in the riveted tank

Riveted tanks (Figure 6.142) may also be protected by a 3D fabric. Such lining complies with all ecologic requirements, regulations for the use of double-bottom tanks and with leakage monitoring system for fuel tanks.

6.2.2   Assessment and repair of underground arterial pipelines

6.2.2.1   Assessment of underground pipelines technical condition

Arterial underground pipelines have become indispensable elements of modern technological civilization. Energy resources such as crude oil and natural gas are carried by pipelines to industrial regions, sometimes thousands of kilometres away from mining places. It is difficult to supervise and constantly monitor underground pipelines, to assess mechanical or corrosion damage that occurs during operation. Emergency leakage, especially in crude oil pipeline that cause product spill are a great hazard to the environment. To diminish hazards on that scale, data on the technical condition of a pipeline related to ageing processes need to be provided at regular intervals.

Currently these are several methods of examining and reporting on the precise technical condition of a pipeline. The most effective is technical condition control using the intelligent plungers – capsules moved inside the pipeline by the pumped media. The capsules are put in and taken out of the pipeline through specially designed chambers located every 80 – 100 km. The intelligent plunger is equipped with sensors [90] [91], which gather information and pass it to the on-board computer. After an inspection trip, the data are analysed by computer and a team of experts, who determine the places in the pipeline where the wall geometry deviates from the standards. Such plungers are the most commonly used control devices that monitor the pipeline walls condition by means of ultrasound, for crude oil pipelines, or magnetic methods to evaluate the geometric section of natural gas pipelines.

The information received from the plunger about pipeline geometry include:
- Mechanical deformation, such as concavities, bulges, bumps, deformations, or ovalization
- Placement and pipe fitting geometry, including information on the damper opening degree
- Root fusions in circumferential welds
- Incorrectly executed circumferential welds (misalignment of joints, different inner diameters of adjoining pipes)
- Deformations, concavities, bulges, and dents in curves; ovalizations
- Persisting material accumulations that decrease the inside pipe diameter, and their location.

A report on the pipeline condition is prepared, summarizing this data. This report determines whether the technical parameters of the pipeline are being maintained, whether it is possible to carry out inspection by ultrasonic or magnetic methods, and whether brush cleanout methods can be used. The report also includes the precise location of all mechanical faults in the pipeline that have been caused during its construction or operation, and also the parameters specifying the size of any deformations discovered.

The plungers that measure the pipe geometry are designed to carry out tests on pipes of a given inner diameter. These devices, used worldwide, are all based on the same principles of operation, and only technical details differ. The mechanical structure of such a device is shown in Figure 6.143. Flips (1) ensure coaxial leading of the capsule in a given pipeline. In the air-tight capsule (2) the on-board computer takes the measurements and registers the results. The ultrasonic dislocation sensors provide signals reflecting deformations of the inside surface of the pipeline. The sensors measure the displacement of the compressed spring measurement lever from the inside surface of the tested pipeline (3).

(a)

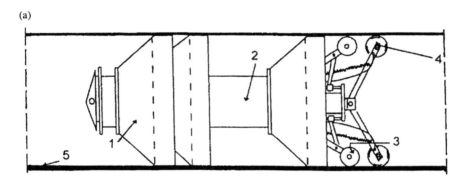

(b)

Figure 6.143   Piston to carry out geometric measurements of the pipeline cross-section: (a) schema, (b) view

These devices have different number of sensors, depending on the pipeline diameter. Plungers to test pipelines of diameter 800 mm most often have eight displacement sensors every 45°, which enables them to calculate four diameters. The coupled levers that co-operate with each measuring sensor have a pair of rollers on the end, which roll along the inside surface of the pipeline being tested. Measurement of dislocation, due to the rollers stiffness, indicates the minimum value of the radius on the pipeline wall arch, over which the rollers have moved. An example of a measurement plunger report printout is shown in Figure 6.144. Correlation of the deformation measurements with the co-ordinate describing capsule displacement along the pipeline is possible using the ushers that measure the distance travelled (4). This measurement is not absolutely precise due to some side movement of the plunger (snaking); moreover, a pipeline tends to be laid in the ground with vertical curves, in relation to the area surface lines, during construction.

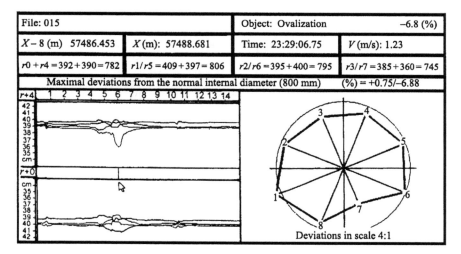

File: 015		Object: Ovalization	−6.8 (%)
$X-8$ (m) 57486.453	X (m): 57488.681	Time: 23:29:06.75	V (m/s): 1.23
$r0 + r4 = 392 + 390 = 782$	$r1/r5 = 409 + 397 = 806$	$r2/r6 = 395 + 400 = 795$	$r3/r7 = 385 + 360 = 745$
Maximal deviations from the normal internal diameter (800 mm)		(%) = +0.75/−6.88	

Figure 6.144 The measuring piston printout describing the place where the pipeline cross-section has been ovalized by 6.8 per cent

Testing of the pipeline walls should be preceded by a geometrical test [91]. If the pipeline has already been tested, it is possible to replace this test with a cleanout using a shield with diameter smaller by about 6 per cent than the pipeline's inner diameter. Depending on the testing device used, the admissible pipe diameter reduction that does not pose a danger to such devices is 10 to 15 per cent. Pipeline wall tests should be carried out every two to ten years, depending on the pipeline operating conditions, its strategic importance and the whether it is environment-friendly. Usually the best interval between tests is five years. It is accepted that the new pipelines do not require such tests, because they do not yet have corrosion pits. However, the justification for carrying out such a test on a new pipeline is clear. It can provide the baseline for the accumulating technical documentation and for preparing the so-called zero test. The test serves as a reference point for subsequent tests and makes it possible to compare results.

There are several methods for testing the technical condition of pipeline walls:
- The ultrasonic method
- The magnetic stream method
- The ultrasonic method for detecting cracks
- The eddy currents method for detecting cracks and hardening.

The best and most frequently used method of testing the wall of crude oil or oil product pipelines is the ultrasonic method. A special testing plunger is equipped with ultrasonic sending – receiving heads, located along the plunger circumference and able to carry out measurements of the whole surface of the pipe. Such plungers consist of several modules, connected with each other by means of a universal joint. From one to five modules are used, depending on the diameter and construction of the device (Figure 6.145). Such modular construction is necessary because of the large number of electronic systems to be packed inside. Moreover, the device has to travel around bends easily. During the travel test the pipe wall thickness is measured, and also the distance from the heads to the wall. When the two results are correlated it is possible to determine whether the corrosion damage is from inside or outside the pipe (Figure 6.146).

(a)

(b)

Figure 6.145 Intelligent piston to test the technical condition of the pipeline wall: (a) piston prepared for insertion in the pipeline, (b) detail of a segment

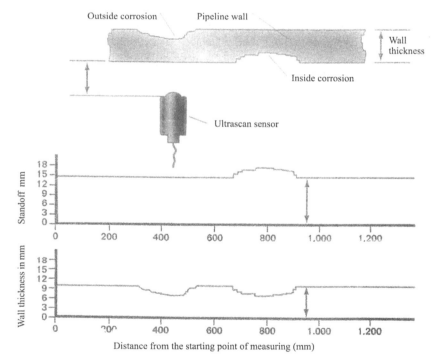

Figure 6.146 Interpretation of measurements by the intelligent piston of the pipe wall thickness

From ground level it is possible to detect damage to the pipeline insulation layer by means of the direct current voltage gradient (DCVG) method. However, to establish whether the insulation damage is related to pipeline damage, it is necessary to dig a strip pit.

6.2.2.2 Repair methods for pipeline damage

The repair methods depend on the size of the operation hazard the damage involves, its character and the affected area. Operation safety considerations are very
important – such as whether the pipeline may be periodically put out of operation.
Regularly used repair methods include the following.

1 Construction of a new pipeline section (Figure 6.147). This method makes it
   possible to repair damage completely, though it is very expensive, troublesome
   and risky. The pipeline has to be temporarily put out of operation and locally
   emptied of its content. Much labour and equipment have to be used. The greatest advantage is that any damage can be rectified, irrespectively of its dimensions and kind, so for a short section with much damage this method may prove
   the most efficient. The amount and degree of damage should always be independently assesses by the pipeline operator.

Figure 6.147 Mounting the new pipeline section in place of the damaged one

Figure 6.148 Mounting of a welding patch on a pipeline

2 Welding of local patches or sleeves over the whole pipeline circumference (Figure 6.148). Welding of patches on a pipeline while in operation is very dangerous. The wall may be burnt through and the medium spilt. Fire or explosion hazards are also very high and the yield point for the heated material may be exceeded. Moreover, some stresses may be introduced to the pipeline that can ultimately lead to emergency state. When the welding works are being done, welds hardened again may become more brittle. Special welding methods must be used, and the welding personnel has to be experienced in such work. The main advantage is that this method can be successfully used when pipeline has leakage occurred through perforation or cracking.

3 ClockSpring Bandages (Figures 6.149 and 6.150). This method involves replacing the wall material loss with filling material and wrapping the pipeline with a composite tape. The principle of this method is to transfer the stresses from the weakened wall via the filler to the composite tape. The strength of a pipe repaired using such a method exceeds the strength of the pipe steel of the top strength, e.g. X80. ClockSpring Bandages are the easiest method and do not require much equipment or manpower. Training of personnel is short (1 – 2 days) and does not require special devices. The most complicated task, which influences the result of the repair, is proper cleaning of the pipe; sandblasting is considered the best method. But however good this method may appear, it cannot provide full leak tightness if the pipe has been perforated on the outside. Also, it cannot be used when the wall loss exceeds 80 per cent of the pipe wall thickness.

Figure 6.149 Repair of a pipeline by the ClockSpring method

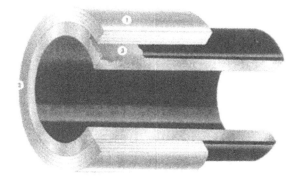

Figure 6.150 Schema of a pipeline repair by the ClockSpring method: 1 a bandage made of the synthetic fibre, 2 filler between the layers, 3 loss-sealing material

4 Epoxy Sleeves PII (Figures 6.151 and 6.152). This kind of sleeve is often used, for they are adaptable to bends, reducers and T-connections. One of their most important features is that they can be used even if the whole inside wall has corroded. This method uses the injection of an epoxy composition between the product pipe and the centrally placed steel sleeve. The method is more expensive compared the use ClockSprings. It can also be modified in practice by using steel belts fastened at the ends of the sleeve to centralize it and to stop the epoxy composite from leaking out.

Figure 6.151 Repair of the pipeline by the Epoxy Sleeve method

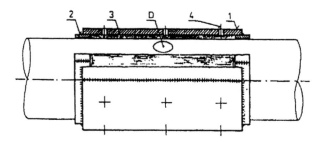

Figure 6.152   Alternative method of repair by a steel-epoxy sleeve: 1 outer steel sleeve,
2 side belts establishing the steel sleeve, 3 filling synthetic resin, 4 vent hole
D pipeline damage

5 Welded or twisted pipes. In this method special pipes are laid on the damaged
   pipeline. They consist of two twisted halves, of a shape that fits the damaged
   area. This method is used when there is pipeline leakage, and can be used not
   only on straight sections but also on any pipeline section (collars, T-connections, bends). The only condition is the performed pipe must secure leak-tightness. The method is mainly used in product pipelines.

   Until recently maintenance services could use only the first two methods, the
   traditional ones. Technological progress in plastics has helped in devising and
   implementing new methods of pipe repair. No pipeline operation stoppages, no
   welding work and no interference to material structure of a product pipeline may
   be required in such methods. Due such advantages, these methods are considered
   very safe and can give results as good as those of the traditional methods.

Selection of repair method on the basis of damage

Each repair method has its drawbacks and advantages, and slightly different applications. Therefore, none can be declared the best for all applications. The distinc-
tive features of each method suit it to the repair of particular kinds of damage.

   For example, if there is some indication that internal damage may spread, then
the a method chosen should be one that can ensure pipeline leak-tightness, even
in case of total corrosion of the pipe wall. ClockSpring does not guarantee such
leak-tightness, whereas all the other methods could be considered in such a case.

   To repair damage to the circumferential welding, all the above methods may be
used, but it should be borne in mind that near the repair area there will be a transverse welded joint, which may complicate laying the repair element. The weld
cutting through the existing weld joint may negatively influence the structure of
the material. With the ClockSpring method it is necessary to carry out additional
procedures to strengthen the material near the welded joint. Besides, this method
does not ensure axial strengthening, which is very important when the damage is
of large circumference. Other methods may be used but economic criteria, the

accessibility of required equipment and the process engineering needed should all be taken into consideration. The Epoxy sleeves PII method seems particularly attractive due to the fact that it does not involve any interference with the welded internal joint structure, while at the same time ensuring the axial strengthening of pipes. Damage to longitudinal or spiral welds as well as to the outer parts of pipelines may be repaired using any of the methods.

The basic criteria for the assessment of damage to the pipe wall should be the guidelines of the ASME B31G standard. However, this standard does not cover other defects and does not provide any guidelines concerning the method of repair. Therefore, the final decision on the method of repair is up to the pipeline operator. Below are some auxiliary criteria for distinguishing the most characteristic kinds of pipeline wall damage:

External losses Losses of wall thickness over 30 per cent should be repaired, e.g. using ClockSprings. In case of losses from 20 per cent to 30 per cent, the insulation layer should be repaired to stop the corrosion processses. Losses below 20 per cent should be left for consideration at the next check. If the cathode protection instillation is faulty it may mean that a third party has damaged the pipeline. In such cases the insulation layer should be repaired to stop the rapid corrosion process from advancing too rapidly.

Inside losses Thickness losses below 20 per cent should be left for further verification. Losses between 20 per cent to 40 per cent on the circumference, corresponding to the sector of a clock face between 0800 nad 1600 (i.e. the upper part out of the pipe circumference) should also be left for further verification. Losses over 40 per cent along the whole circumference and losses between 20 per cent to 40 per cent corresponding to the sector of a clock face between 1600 and 2000 (i.e. the lower part of the pipe circumference) must be repaired. Losses between 1600 and 2000 are in danger of increasing rapidly because of the gathering of water or other corrosive solutions (unless the pipeline is dry and the medium non-corrosive), so the repair method will have to ensure leaktightness where there is complete wall perforation.

External losses at circumferential welds These are repaired as external welds, but the ClockSpring method is not recommended, due to the lack of axial strengthening, if the damage is over 30 per cent of the wall thickness and includes a pipe section of total circumferential length amounting to 30 per cent of the pipe circumference.

Delaminations and non-metallic inclusions Also called laminations, these are not treated as damage hazardous to pipeline reliability. However, skew lamination touching the wall surface must be repaired by the replacing of a section pipeline. Surface laminations of depths up to 30 per cent from the external side may be ground down to a healthy layer of metal and then repaired as with ordinary external losses.

Cracks If the cracks are neither too deep (up to 30 per cent of the pipe wall thickness) nor too extensive (just single surface cracks), they should be ground down to sound metal and repaired by one of the method used for external losses. In the case of extensive cracks, the faulty pipe section should be replaced. Even if the crack is shallow and single, the mechanical properties of the pipeline will probably not comply with the strength requirements. Therefore, a better solution would be to replace the whole section. In the case of a plain longitudinal crack, of length not less than the diameter and not larger than 0.3 m, the repair can be temporary, made by means of a two-halved sleeve.Ultimately however, pipe sections with such cracks should be replaced, as the whole pipeline is likely to have inadequate mechanical properties or material defects.

6.2.2.3   Replacement of damaged pipeline sections without shutting down operations

Replacement of a damaged pipeline section without putting the pipeline out of operation is technically possible with the use of special equipment, whose primary element is a machine that can cut out holes in an operating pipeline. The sequence of procedures for this method is shown in Figure 6.153. The first step is to pad weld on each end of the operating pipeline the connectors to the by-pass pipeline, and identical connectors to close off the flow (Figure 6.153(a)). These connectors consist of the collars with a two-part sleeve to grip the pipeline (Figure 6.154) After leaktightness pressure and weld strength tests have been carried out on the connectors, the plate gates are mounted and the special machine for cutting out the hole by the Williams's method (Figure 6.153(b)) is attached. This machine (Figure 6.155) is equipped with a set of milling cutters, drills and a gripping device that can cut out a pipe segment, hold it and move it to the outside the pipe, after the plate gate has been closed and the cutting machine has been disassembled.

The holes in the external connectors are cut first, and the by-pass pipeline is fitted. When the plate gates are opened the medium will flow in the main pipeline

(a)   (b)

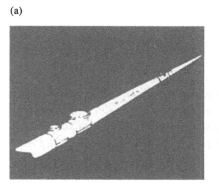
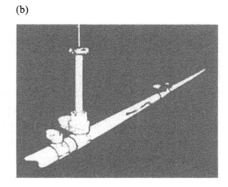

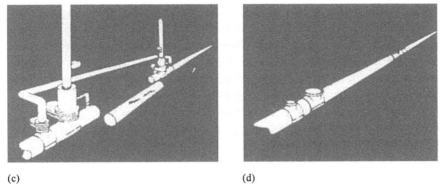

(c)   (d)

Figure 6.153 Subsequent phases of damaged pipeline section replacement while still in operation: (a) pad-welding of connectors (see Fig.154), (b) cutting out the pipeline wall inside the connectors, (c) mounting of a by-pass and after closing the flow in the main pipeline, cutting out the damaged segment, (d) dismantling the by-pass and stopple plugging

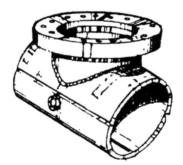

Figure 6.154 Connector pad-welded onto the pipeline

Figure 6.155 The fittings to cut out the holes

and the by-pass at the same time. The next step is to transfer the cutting machine to the external connectors and begin drilling the holes to install the stopple pluggs, which close the flow in the pipeline. From then on the pipeline is plugged in two places and the damaged section can be replaced (Figure 6.153(c)). After the repair work has been finished the stopple pluggs are dismantled and the flow through the repaired pipeline section can be restored. Then the by-pass pipeline and the plate gates can also be dismantled (Figure 6.153(d)). This is possible because of the design of plugs of the Lock-O-Ring type. On the collars there are also regular pipe plate stoppers with seals.

6.2.3 Assessment and repair of gas pipelines

6.2.3.1 Cleaning of gas pipelines

During the operation of pipeline systems it is necessary to clean them regularly. The purpose of cleaning is to remove various kinds of impurities. Currently, these impurities include: oil that has escaped from technical devices, leftovers from technical or repair work, such as dust and in particular metal impurities in the form of welding electrodes or metal facings from grinding off or blasting pipeline parts. Metal facings represent a special problem because they can cause signal distortion during inspections.

Nowadays, only cleaning device that have been specifically designed for the removal of liquid and metal impurities are used in pipeline systems. Calibration plates can be used for determining pipeline capacity during cleaning, though profile inspection devices can determine capacity more accurately (Figure 6.156).

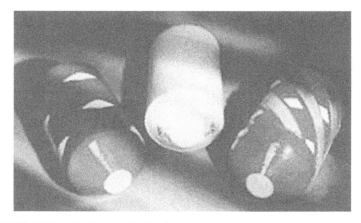

Figure 6.156 Device for cleaning the pipeline

Figure 6.157 Cleaning of a pipeline system

Figure 6.158 sets out a general procedure for cleaning pipeline systems. This procedure is appropriate only for a pipeline that has never been cleaned since it began operations or for which data on past cleaning is not available or is unreliable.

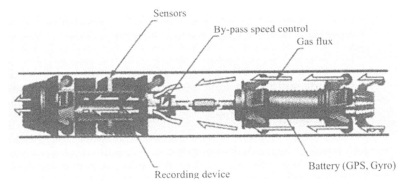

Figure 6.158 Standard inspection device

Basic polyurethane foam cleaning devices

These are used as the first step in examining pipeline capacity. It is made from an elastic foam material and along its circumference there is a polyurethane layer to decrease friction and wear. This device is a a kind of searcher and in cases of partial pipeline capacity its destruction (and disposal) can be very expensive. Such devices have a capacity of 40 per cent of the nominal cross-section of the pipeline to determine capacity more accurately a calibration plate with the size of the minimum capacity of the cleaning device is used with the more sophisticated kinds of cleaning devices, to be described next.

Devices designed for removal of liquid impurities

The use of this kind of devices represents the second step and after their use the pipeline should be free of a substantial proportion of liquid impurities.

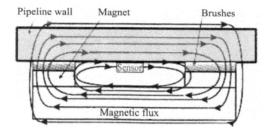

Figure 6.159 Schema for a magnetic cleaning device

Devices designed for removal of metal and non-metal impurities from pipeline wall

If a corrosion and profile inspection (examination) is to be carried out, the pipeline walls should be cleaned due to improve the quality of the data to be inspected.

Devices designed for removal of metal and non-metal impurities from the pipelines

Because metal impurities endanger the sensors of inspection devices they represent a major damage hazard that can result in data incompleteness. Therefore proper cleaning of them from pipeline is essential. For the removal of metal impurities it is necessary to use magnetic devices. There are various traditional cleaning devices which differ only in the magnetic segment fitted in their centre. The Slovak gas industry uses magnetic cleaning devices of their own design, which were used in

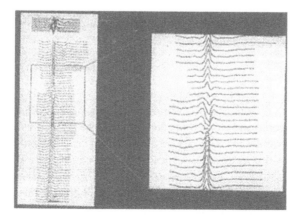

Figure 6.160 Data from a magnetic cleaning device

conjunction with the latest technological elements yield outputs exceeding by many times that of similar devices used around the world.

Each device is equipped with a two-way radio transmitter that makes it possible to determine precisely the position of the device in the pipeline. Such checking is essential, particularly in dealing with streching of the cleaning device in the pipeline chamber (Figure 6.160).

Cleaning of the pipeline is carried out under normal operational conditions using the Start and Receiving chambers. During operation the device is monitored by checking pressures with TU or by land microphones.

6.2.3.2 Assessment and its application during operation of a transit system

Internal inspection of pipelines is considered the first method of determining the operational reliability.

The inspection device used in the assessment of pipelines transporting material in a gaseous state uses propagation of changing magnetic fields for the measurements (Figure 6.161). The magnetic field is created by the set of permanent magnets along the whole circumference of inspection device, from which the magnetic flux is carried via steel brushes to the pipeline wall. As the polarization of the permanent magnets alternates, a circular magnetic field is created close to the pipeline wall and the body of the inspection device. As the inspection device moves a homogeneous flow of magnetic field is created in the pipeline wall, the intensity of which is strong enough so that changes in the homogeneity of the magnetic field in the pipeline wall caused by non-magnetic defects or additional metal elements in the pipeline wall or its vicinity will be detected by a number of sensitive sensors, which measure their distances and lengths simultaneously (Figure 6.162).

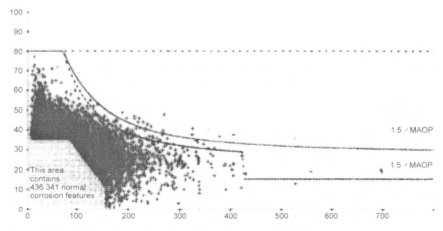

Figure 6.161   Data on the prolongation of changing magnetic fields

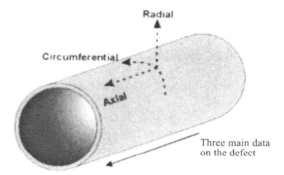

Figure 6.162   Dimensions of measurement

Non-magnetic defects include:
- Pit and planar corrosion
- Cracks and ruptures
- Doubling and overlapping
- Insulation joints
- Technical and constructional interstries on armatures and branches.

Additional metal elements include:
- Roots and covering of all kinds of welds
- Changes in wall thickness
- Welded-on pieces of anticorrosive protection
- Forming elements and branches
- Guards
- Supports and girders
- Wall slumps
- Foreign metal objects in the vicinity of the pipeline.

Signals from the sensor related to magnetic field changes and measurements of distance are passed through an amplifier for further processing via filters and transducers. All signals in modern inspection devices are translated into digital form, which is more suitable for processing and less demanding on memory. Recording on magnetic tape was used in the first memory systems, but such a system is space-demanding and its memory capacity does not meet the requirements of single-run inspections.

Very reliable, small and high capacity recording devices include digital audio tape (DAT), EPROM memories, and hard disk drives (HDD).

For gas pipelines only devices with high resolution are used with threshold sensitivities at the required accuracy of approximately 5 per cent of the nominal thickness of the pipeline wall. This sensitivity is sufficient for deciding the strategy for pipeline operation, because defects up to 17 – 20 per cent of the nominal thickness of the wall have no substantial influence on their operation.

Evaluation of the data is done by 'spreading' the signal obtained for the whole surface of the pipeline by a rose arrangement of sensors and comparing the developing curves using etalons. In the analysis, technical workers use their experience and imagination together with records of distances, depths and lengths of defects calculated by correlation coefficients after data filtration. For this analysis, maps of the location of the pipeline in the surrounding area, drawn at a scale of 1:5000 are also useful. These maps are particularly useful for identifying the arches, crossroads or other communications (Figure 6.163).

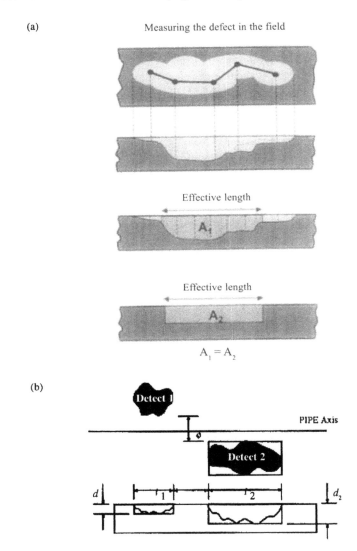

Figure 6.163   Pipeline location maps

In general, inspection devices that work on the principle of magnetic field variations in the pipeline wall are used in gas pipelines. With gas pipelines, inspection devices that work via ultrasound are difficult to use because they need a liquid medium between the pipeline wall and sensor.

Recently, inspection devices have been equipped with GPS measuring systems which enables accurate mapping of the pipeline on the ground and the generation of 3D models. This makes field work more accurate and less expensive.

Inspection companies will process the data and tabulate it according to the nature, size and distances of defects. Information provided includes:
- The distance of objects detected from the start of the section inspected
- Inspection report for the section
- A description of detected objects.

Which criteria are to use in the evaluation of defects is to a great extent determined by the company that controls the pipeline. Only occasionally are there cases where the criteria are determined by government regulation or technical standard.

The American standard ASME B31G is generally accepted around the world as the basis for evaluation of material losses from the pipeline walls caused by external or internal corrosion. Essentially this involves plotting the defect on graph. On the horizontal axis is the length of the defect, and on the vertical axis the depth expressed as a percentage of the nominal thickness of the pipeline wall. The limit of availability is determined by the curve for the coefficient ERF = 1 (broken). The EFR coefficient (factor of repair) is calculated from the basic geometric dimensions of the defect and the operational and constructional pressures of the pipeline being inspected. The ERF coefficient is calculated for every corrosion loss. Where there are more serious defects, the use of simple lengths and depths is dangerous. In such cases the R-STRENG and LAPA methods are used. These use the so called 'bottom of the river' technique: the accurate mapping of the defect profile. This technique is a demanding form of defect evaluation, and the cost of evaluating each defect is 10 – 20 times higher than simply determining the defect length and its maximum depth. Therefore transformation rules have been prepared for converting the latter to the R-STRENG method. The reduced coefficient of repair curve ERF = 1 is displayed as a continuous curve just above the curve ASME B 31 G ERF = 1

Corrosion thus identified can be divided into two kinds:
- Pit corrosion: when the axial length is not more that three times the nominal thickness of the pipeline wall
- Planar corrosion: when the depth is more than three times the nominal thickness of the pipeline wall.

6.2.3.3 Separating pipelines on a pipeline bridge

During assessing inspections it was found that on the pipeline bridge for Transit Gas Pipeline TGP I and TGP II Js 1200 the pipelines bridge had merged together at the crown of the arch (Figure 6.164), making it difficult to inspect and maintain them. Close examination showed that the pipelines were originally erected with relatively large alignment deviations.

Figure 6.164   A view of the overlapped of the pipelines in the crown of the arch

Pipeline TGP II had a maximum horizontal deviation of 220 mm from the theoretical position of the horizontal track and pipeline TGP I about 100 mm. From a theoretical analysis it was obvious that the pipelines had been influenced by temperature effects. Due to height and orientation of the pipelines, separation by symmetrical pushing was impossible. Given this fact, a method of re-positioning the pipelines was agreed that involved moving them from their present horizontal track into the required position using a temporary device.

During the preparatory work the original bearing saddles at the locations where the pipelines were to be separated and other saddles were modified for shifting in the horizontal direction. While removing the original saddles, the horizontal and vertical movements of the pipeline were monitored. Before the relocation of the pipeline began, its position was secured against other movements and additional pulling devices were fitted on the bridge (Figure 6.165). These consisted of a half-ring clamp (set on a pliable base), draw bars and a movable binder. Where the pipeline was to be moved, a pliable base was put under the steel half-ring clamp. After setting and fixing the base, the ring clamp was fitted, to which the draw bars – threaded bars with a circular cross-section – were connected. Gradually, the various elements of vertical brackets (Figure 6.166) of the auxiliary device were inserted in the gaps between the flat profiles of the adjacent main girder. An auxiliary horizontal bracket was connected to the vertical brackets to support the hydraulic press. The free ends of the draw bars fitted with threads were anchored in the movable binder. After completing the erection procedure, the hydraulic presses were set for inducing the appropriate horizontal loadings.

Figure 6.165 The separating device

Figure 6.166 Setting the separating device on the upper girder

When the preparatory work was completed and working order of the proposed devices was checked, pipeline TGP II was depressurized. After the pressure had fallen to 4.5 MPa, a check measurement was made, which showed that in the horizontal and longitudinal movements there was a flexible distribution of stresses towards the linear section of the pipelines, on both the sides. Given the finding that any additional longitudinal stresses did not originate in the pipeline body, depressurizing was stopped at the specified value of 4.5 MPa for the internal pressure in the pipeline. In this state the moving of pipeline of TGP II was begun using the pulling devices. The supports for the pipeline on the both sides of the bridge were gradually released from the crown of the arch. In two places (Figure 6.167) pipeline TGP II was moved in small steps into its new position by means of hydraulic presses, with the horizontal loading increasing by 20.0 kN increments.

Figure 6.167   Locations on the crown of the arch where the separating device were fixed

Because such rectification work on a pipeline in operation that had not previously been done in Slovakia, it was necessary to monitor the pipeline stresses at each loading stage. In the preparatory work a tensiometric monitoring system was installed on pipeline TGP II. Stresses were checked by this system during depressurizing, rectification and returning to full operation. The whole procedure was controlled so that the pipe was in a flexible condition at every moment in the separation to ensure the resulting stresses kept to a minimum.

The maximum measured input stresses immediately after completion of three separation cycles was 62.17 MPa. On the theoretical assumptions that the pipes would behave rigidly, stresses up to 110 MPa were expected. With pipelines TGP I and TGP II, which were originally tight together at the crown of the arch, the gradual separation created a gap with a minimum width of 229 mm (Figure 6.168) at the crown of the arch. In the horizontal movement there was an elastic distribution of stresses due to longitudinal movement towards the linear sections of the pipelines at both ends of the bridge. Therefore up to a horizontal separation of 229 mm was reached, significant additional longitudinal stresses had not developed. At that value for the horizontal separation, the scope for longitudinal deformability of the pipelines was exhausted and at further horizontal loading the total stresses would not exceed the admissible values.

Figure 6.168   A view of the pipeline after moving it into the new position

When pressurizing to full operating condition, the pipeline was cushioned so that the resulting measured values for input stresses did not exceed 20 MPa. During the entire rectification work the operational reliability and safety of the pipeline system on the bridging had been secured by an innovative technical procedure.

6.2.3.4 Repairs to faults in anchor blocks

In anchor and brake blocks of river bridges there is often excessive corrosion losses in the basic pipeline material at friction clamping rings and pipeline passage guards placed in anchor blocks. Before such faults are repaired, the friction clamping rings must be completely replaced and additional anchor blocks built and coupled to the existing anchor blocks, which may also eventually be replaced by new ones.

Two pipelines TGP I Js 1200 mm and TGP II Js 1200 mm crossed the river Údoč via a pipeline bridge. In the anchor blocks of TGP II corrosion losses of the basic pipeline material was detected on the both sides. In the repair, the original anchor blocks had to be replaced by new ones. The anchor blocks were located below the level of the adjacent ground and after studying the available documentation, new locations for the replacement of anchor blocks were chosen (Figure 6.169) Due to the arrangement of the pipelines and the original anchor blocks, the new blocks were placed in the space between the outfall of the pipeline from the bridge and the old anchor blocks.

Figure 6.169   A view of the old anchor block

In the proposal for the repairing the anchor blocks of the bridge, various geotechnical solutions and alternative design for additional steel load-bearing structures were submitted. For the geotechnics, the basic archive documentation was missing; this could be replaced by appropriate engineering-geological research. The loads to be carried by the anchor blocks was determined by interactive feedback calculations. This proposal generated savings by using technologies that made possible a constructionally simple, reliable and cheap solution.

The optimal derived solution required ten drilled piles on one block. Given the results of the geotechnical calculations, the steel structure for the anchor blocks was designed for a vertical loading of $V = 2646$ kN and a horizontal loading of $H = 1856$ kN. The anchor block for pipeline TGP II was fixed on to steel load-bearing framework (Figure 6.170). This load-bearing structure had a designed width of 3.4 m and was connected to the base structure by anchor bolts and shear stops. The axial scheme of the steel structure follows the shape and position of the piles. The framework was made from HEB rolled profiles, with the vertical loading from the pipeline being carried by four upper binders of HEB 500.

Figure 6.170   The steel load-bearing structure of the new anchor block

The horizontal loadings from the pipeline are transferred to the load-bearing framework by means of four two-part friction clamping rings (Figure 6.171). The faces of the clamping rings rested against the side of supporting brackets of the steel framework, which prevented it from shifting onto the load-bearing structure of the bridge. After constructing the steel structure it was covered with concrete.

Figure 6.171   The new friction clamping rings

When anchor blocks are replaced, the local conditions in the pipeline supports will change, resulting in the development of additional stresses from girder activity of the pipeline. During the repair, using tensiometric measurements, the stress in the body of pipeline TGP II was monitored at every stage of construction (normal operational condition, depressurizing for installation of the new anchor blocks, activation of the new anchor blocks, re-pressuring). After the repairs had been completed, the monitoring system was retained for long-term monitoring of stresses and deformations in the repaired pipeline.

Figure 6.172   Detail of the supports for the friction clamping rings against the supporting brackets

In the anchor block for pipeline TGP II in the bridge over the river Váh, corrosion losses were found on side B of the basic pipeline material. The original anchor blocks were below the level of adjacent ground. After examining the ground and studying the documents provided, the position of a new anchor block was chosen. Given the arrangement of the pipelines and location of the original anchor blocks, it was decided that the new block could be located in the space between the opening for the pipeline from the bridge and the old anchor block.

The starting point for the feedback calculation was the requirement that all the horizontal forces be taken by the new anchor block. A reinforced concrete gravitation anchor block was designed (Figure 6.173), and set in front of the existing block leaving a space of 2.0 m. The new anchor block is able to transfer a horizontal force of: H_{max} = 1302.0 kN, with the force being transferred to the anchor block by a steel stop. The stop consisted of a four-part clamping ring of 16 mm thick plate (Figure 6.174).

Figure 6.173   A view of the new reinforced-concrete gravitation block

Figure 6.174   The four-part friction clamping ring

During construction the stress conditions in the pipeline were monitored continuously, so that the operational reliability of the pipeline guidance was secured. The effect of the repairs to the anchor block on side B of the bridge was monitored by long-term tensiometric measurement. This monitoring system was set up and activated before depressurizing the pipeline and beginning the replacement of the anchor block. Comparison of the theoretical values for the assumed stresses and results of measurements showed a good correspondence. It was therefore possible to state that the maximum input stresses on the pipeline did not exceed a value of 15 Mpa, so that over the whole period of the replacement of the block, operation reliability was secured.

6.2.3.5 Elimination of a short circuit in the cathode protection of pipeline guards

The pipelines in a culvert under a road pass through steel tubular guards. Due to movement in the pipeline and its surroundings, there can be metal contact between the pipeline and the guard causing a short circuit of the cathode protection.

Breaking the contact between the guard and the pipeline was done by raising the pipeline and adjusting its position in the guard. The required maximum elevation of the pipeline was 50 mm and during lifting a normal operational pressure in the pipeline of 6.4 MPa was assumed. Before starting the repair, the strength of the pipeline in the vicinity was assessed at the internal operational gas pressure.

Figure 6.175 A view of the lifting device in the excavation

In some sections of the guard the short circuit was removed by raising the pipeline in the guard using a lifting device. Before lifting, the pipeline was uncovered at both ends of the guard for a length ranging from 20 to 25 m (Figure 6.175). During digging in the vicinity of the supporting points for the lifting devices, the guard was modified so that the load distribution panels could rest on a prepared ground layer. Before laying the load panels (Figure 6.176), the ground was dug manually and prepared. The supporting points in the longitudinal direction for the hydraulic presses to raise the pipeline were located at distances of 1.0 to 2.0 m from the guard ends.

Figure 6.176 Placement of the lifting device on the load distribution panels

The pipeline was lifted using a suspension draw system. At the half clamping rings, a flexible base was fitted between the pipeline covering and the plate of clamping ring. The suspension draw system consisted of a half clamping ring made from hot-formed P 16.300 plate and ϕ 48 vertical bars. The bearing binder was made from two HE 300 B girders and was laid on supporting columns. The draw bar system was installed and the formed bases welded where the 200 kN hydraulic presses were located. After emplacing the presses and the 2 × U200 movable binders, the system to lift the pipeline was ready (Figure 6.177).

Figure 6.177   A view of the rigid, movable binder and lifting presses

During lifting, the stress in the pipeline covering was monitored tensiometrically. The sensors were placed in the lifting area and where the pipeline entered the ground. At the same time, deformations in vicinity of the guard ends were also monitored. In all repairs under operational pressures, the safety rules for working in vicinity of a gas pipeline must be followed.

6.2.3.6 Lowering flooded pipelines

During floods pipelines are often flooded. At one locality, the pipeline was flooded above ground for a length of 92.0 m (Figure 6.178) and insufficient soil covering was left over a section of about 250 m. At the same time the pipeline was off-centred from its horizontal axis by about 1200 mm and approximately 700 mm above the ground. Tracks showing that agricultural machinery had ridden over the flooded pipeline were assessed but no cracking was detected by magnetic tests. Scratch traces were removed by grinding to a maximum depth of 1.0 mm. The original damaged insulation was removed and the pipeline was cleaned by sand blasting. The length to be repaired, of approximately 150 m, was loaded with standard reinforced-concrete loading saddles. The density of the arrangement of loading saddles, as well as the distances between supports of the pipeline during lowering were determined from static calculations. The pipeline insulation is

Figure 6.178   The flooded pipeline

protected by 10 mm geotextile layers. This soft base can moderate local effects on the pipeline at places of contact with the rigid saddles.

The pipeline was temporarily supported by ground supports at an axial distance of 36.0 m and with base width of 6.0 m before lowering. The first support was located at the centre of the rising arch. Due to the large deformation of the pipeline, the length of excavation for the edge sections, as well as the width of excavation, had to be sufficient to enable natural laying of the pipeline together with the necessary covering over the loading saddles. As the tubes could not move under the loading of their own weight at the given unloading, the pipeline lowered by gradually under cutting ground supports (Figure 6.179) while the tube was for some moments non-load-bearing over a length of about 72 m. The whole procedure was controlled so that at every moment of lowering the pipe would be in a flexible condition. While the pipeline was being lowered, the internal gas over-pressure was 3.0 MPa.

Figure 6.179 Undercutting the ground supports

To measure the normal stresses on the φ 1400.16 pipe body of the pipeline while it was lowered into the excavation, electric rheostatic tensiometry was proposed and subsequently used. To set the tensiometers, the places with maximum expected values of stresses generated during the pipeline were chosen. Hottinger BM foil tensometers were used.

According to theory, as well as calculations carried out at places of lowering, the expected maximum stresses during lowering should range from 85.6 MPa for a favourable case, while when the support was cut away at once a value of 165.0 MPa would be reached. If the pipe remained in the air at the place of the support, a value of 320 MPa would be reached. Therefore a gradual removal of ground support was chosen with a maximum stress of 3.0 MPa in the pipeline over particular time intervals while under a gas pressure was 215.6 MPa. This proce-dure, as well as monitoring of the pipeline lowering into the excavation, secured the safety of the surroundings and operational safety of the pipeline.

When the pipeline had been lowered to the required depth, the reinforced-con-crete loading saddles DN 1400 of 22.2 kN (Figure 6.180) were laid on the pipe-line. The insulation under the saddles was protected by 10.0 mm geotextile layers. The loading saddles are arranged at axial distances of 1200 mm each, i.e. at a block width of 600 mm, the width of the distance between the saddles is 600 mm. During lowering the pipeline with the fourth line DN 1400 in the area, the values

Figure 6.180   Emplacement of the reinforced-concrete loading saddles

of the stresses measured experimentally by tensiometers did not exceed theoretical values from the self-weight of the pipeline. At all times during lowering the pipeline was in an elastic condition and the resulting stresses did not reach the limit of elasticity for the pipeline material. The procedure used for laying the pipeline in this case has been applied in a modified form in similar repairs and shifts of pipeline in the places and under other conditions.

6.2.3.7   Adaptations in slippage areas

Transit gas pipelines in many localities are subject to slippage. The slippage conditions in some localities with regard to pipelines may be evaluated as critical and require prompt solution. By the time the stabilizing of slide slopes in such localities has been carried out, the additional stresses in the pipeline cross-section assumed to have been caused by the slide need to be evaluated and sites for monitoring other effects of the slide on pipeline should be chosen. The arrangement of tensiometric detectors in individual probes along the length of the pipeline in the monitored section, as well as the arrangement of sensors around the pipeline circumference in individual cuts is devised on the basis of a theoretical analysis for the given locality. The effects of the slide on pipeline stress is monitored by long-term tensiometric measurement. During the monitoring of the pipeline in the slippage area, total pipeline stress is checked so that the operational reliability of the pipeline can be secured.

Before the proposal for monitoring the slide in the discussed below locality was devised, two examinations of the slippage area had been carried out, in the autumn of 1999 and February 2000. During the first examination of the third pipeline and its surroundings in November 1999, the pipeline was uncovered in two test pits. In the first larger test pit away from the road, which was about 8.0 m long, the pipeline was repaired (Figure 6.181). During this examination the water level reached the lower parts of the pipeline. The uncovered pipeline was slightly deformed towards the foot of the slope. During the second examination in February 2000, the first larger test pit was flooded to the extent that the pipeline was below the water level. Over a section approximately 350 m long between the road and assumed stone bedrock the route of the pipeline was marked. On the basis of the results and calculated values for probable deformations, a calculation model for determining the additional stresses in the pipeline cross-section was devised. The stresses were determined using two methods: membrane band theory and solid fibre theory. The theoretical values obtained for normal stresses caused by the slide did not exceed 38 MPa. From this theoretical analysis of the stresses it was apparent that the pipeline was designed using the method of admissible stresses; so there were no reserves for the additional stresses resulting from the change in the edge conditions of the pipeline emplacement.

Figure 6.181   The condition of the pipeline in the excavated test pit

To measure changes of stress, four open test pits were proposed in the section between the road and assumed stone bedrock. Tensiometric sensors around the pipeline circumference were installed in the test pits so that the initial measurement state and null reading could be defined.

When another change in the pipeline emplacement occurs, i.e. the pipeline is dug out, also the dynamic conditions and normal stresses are changed, and these changes are also recorded. The sensors in the test pits monitored the situation in the original excavation during the subsequent depressurizing and repair works. While depressurizing the pipeline, a value of $\Delta \sigma = 338.0$ MPa was measured in the cross sensor, which exceeded the theoretical value by approximately 50 MPa. A defective weld in the pipeline was repaired in this case by replacing a section of pipe. The cutting and subsequent insertion of the new section is shown in Figures 6.182 and 6.183 . After cutting and subsequent stabilization of the pipeline a pressure of 40 MPa was released over the given section. The pipeline was now behaving elastically and the components of the additional longitudinal stresses caused by slide were eliminated.

Figure 6.182   The cut pipeline

Figure 6.183   Insertion of the new pipeline section

After the repair, the whole section was refilled and the condition of pipeline was monitored. It is recommended that further periodic monitoring be carried out four times a year, or after each change in bedrock properties due to excessive rainfall. If there has been a change in longitudinal stresses of the order of 30 MPa it is necessary to identify the cause of this change and devise a method to adjust the stresses in the pipeline.

6.2.4. Asessment and repair of steel chimneys

6.2.4.1 Construction and operating damage specific to steel chimneys

The chimneys can be divided into two kinds according to their functions:
- Smoke chimneys: used for releasing the exhaust fumes and dust produced by the contribution of carbon, gas or fuel oil in the boiler and industrial furnaces
- Discharge chimneys: used for removing gases and other air pollution generated at work places and in some industrial processes.

Steel is often used in chimneys because it allows uncomplicated construction, quick erection and easy dismantling after its technical operating life has expired.

Steel chimneys use four basic constructional designs (Figure 6.184). Detached chimneys constructed as vertical cantilevers sat on a base foundation are the most common, and can be up to 120 m high. The chimneys stabilized with guys require a large area in order to provide the best angle of inclination for (45 – 60°) the guys. Chimneys of this type, as well as the chimneys supported by a tripod, have heights of 45 – 70 m. The greatest height (up to 300 m) can be obtained when the chimney shaft is supported by a 3D lattice tower.

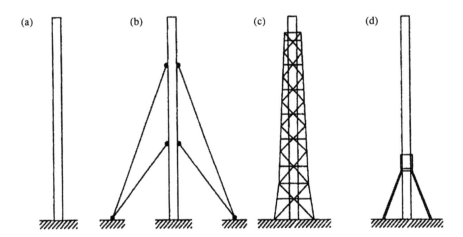

Figure 6.184   Types of steel chimney: (a) detached, (b) with guys, (c) inside a lattice tower, (d) supported by means of a tripod

The chimneys releasing hot and highly aggressive gases have an inside lining of gunite, low-alloy steel or stainless steel. Two-halved chimneys consist of two steel pipes of different diameters concentrically located with thermal insulation between the pipes. The largest group of the chimneys both constructed and under construction are the single detached chimneys. However, their durability is questionable because of high failure frequency due to faults, construction and design defects, or improper operational use.

Although the material deficiences are rarely mentioned as the chief reasons for failures or defects, it should be noted that the design and construction shortcomings are often to blame. The selection of the correct steel grade and type, electrodes and welding techniques, screw grades, kind of insulating material in field collars etc. is important. Many a time changes were introduced arbitrarily: the grade steel was changed to an the unalloyed kind, heat-resisting steel was used instead of acid-resistant steel, auxiliary materials or welding techniques did not comply with the project's requirements. Another controversial issue is the materials used for the surface anti-corrosion insulation. However, though the letter this issue is important, it is not the primary concern relevant to this book.

Severe and extensive corrosive damage in steel chimney shafts can be the ground for the judgement failure is principally caused by steel corrosion. Medium losses in the plate thickness of the single-shell steel chimneys used in the local and industrial heating plants, calculated from 140 examined chimneys, amounted to 0.3 mm per annum in the middle segments, 0.4 – 0.5 mm per annum in the upper segments, and up to 0.65 mm per annum in joints and vicinity of fasteners areas [93]. In some steel chimney shafts inspected, greater or lesser corrosive damage was detected, the greater damage occurring close to shaft collars and the supporting ring in the chimneys stabilized by tripods. Some chimneys were extensively damaged by corrosion of plates and shaft segment joints; numerous perforations, cracks and local deformations, etc. were plainly visible.

Corrosive damage to chimney shafts did not vary at collar joints or the height above the gas inflow to the chimney. Variable degrees of damage may be the result of the prevailing wind direction, and the lowered temperature of the surface of the steel plate in the windward direction. The region subject to most corrosive damage are pipe segments that are below the dew point, because there the steam and acids in the fumes will condense on the inner surface of the chimney shaft. Otherwise the corrosive damage to chimney shafts tend to occur at random and therefore in the detailed analyses of damaged chimneys and their capacity they should be treated accordingly.

Improper manufacture and erection of steel chimneys may also accelerate corrosion damage (e.g. the faulty field welds on shaft segments or tight deformed collars, changing the specified thickness or grade of steel, faulty application of the specified surface protection). Mistakes and negligence during construction and erection are the most frequent reasons, cited by experts for the unsatisfactory technical condition of some steel chimneys.

Corrosion of steel chimneys is a function of the operating conditions, which are often adverse the basic design. Particularly hazardous can be alterations to chimney operational techniques that lead, among other things, to changes in temperature and pollutants in exhaust gases, and frequent operational stoppages. Current environmental protection regulations require the fitting of fume desulphurization devices, which result in lowering temperature of the exhaust gas and increasing their humidity through the desulphurization process. This is a very grave problem for existing industrial chimneys, notably power station chimneys. With steel chimneys, this problem mainly concerns the exhaust gas duct located inside tall multi-duct reinforced steel chimneys.

Vibration in steel chimneys may be caused by wind gusts (vibration orthogonal to the wind action, at relatively low wind velocity) and mechanical excitation by, for instance, operating equipment or dedusting devices. The most common cause of excessive vibration in detached steel chimneys, is periodic and alternating excitation by turbulent vortices on the left side of the cylindrical shaft surface. Because these vibrations depend on the frequency in the free vibrations of the chimney, change in the basic dynamic characteristics of the structure as erected and operated that can lead to unforeseen resonant vibrations should be taken into consideration such as (weight and rigidity). Such vibrations can also be caused by other factors: the location of ladders and walkways, aerials of mobile phone network, nearby buildings [92].

A separate kind of reason for corrosion damage to steel chimneys is excessive fatigue of materials due to the rapidity of steel corrosion and excessive chimney shaft vibrations unexpected by the designers. Rapid corrosion increases the occurrence of fatigue cracks. In addition to the high-cycle fatigue steel chimneys also suffer problems from low-cycle fatigue. Factors that affect basic durability, such as corrosion and fatigue, are included in the requirements of standards DIN 4233 [95] and CICND [96].

6.2.4.2 Periodic checks and examination of the technical condition of chimneys

The technical condition of steel chimneys should be checked periodically because violent damage – such as chimney collapse – is extremely hazardous to the environment. Usually the first check on technical condition is required after four years of operational use, with subsequent checks scheduled by the inspector in the post-check guidelines. All chimney elements should be subject to a condition check: the shaft, joints, foundations, and the supporting structure (supporting tower, tripod, guys) if it part of the chimney structure. Tests on the shaft should include the appraisal of the steel wall thickness, degree of corrosive wear and the amount of cracking in the lining.

Shaft wall thickness measurements should be carried out by means of non-invasive methods (using an ultrasonic thickness gauge), along the whole shaft and paying special attention to segment joints and supporting bearings. Measurement points shall be located every 0.5 – 1.0 m along the shaft, and at least three measurements should be made at each point. At collar connections (where more extensive corrosion tends to occur) measurements should be made directly above and below the collars (between the ribs) and directly above and below the ribs. From three to six measurements should be made above and below the coupling. If there is a large scatter of results, the number of measurements points should be increased. The measurement precision should be to 0.1 mm. The mean values should be used for wall thickness. On the basis of the measurements obtained, the shaft corrosion rate from the beginning of operational use, may be estimated, as well as the annual losses, which together enable preparation of a structure durability forecast.

After referring to the design documentation a survey of the welds should be carried out in accordance with the applicable standard. If the documentation states that the welds are faulty, of poor quality or have cracks, then weld checks should be carried out. Mainly it will be the horizontal welds in the shaft sections mast subject to stress that will be checked, by random tests from 30 to 50 per cent of the sections. These weld tests should use penetration, ultrasonic, or radiographic methods in accordance with the applicable welding standards. The defective proportion of chimney shaft welds should amount to at least class C, according to the EN-25817 standard. The screws, the most important element of chimneys, should be checked to:

- Determine their class and diameter in comparison with the basic design specification
- Estimate faults in the couplings (e.g. use of arbors are too short, corrosion of screws, wrong washers or screw caps, insufficient tightening).

If the design specified that screws of high strength were to be used, but there is no indication of what class of screws were actually used in construction, the strength of screws selected at random should be tested (at least three tests for each diameter).

The technical condition of the chimney lining may be assessed during operation using a thermo-visual camera, and when the chimney is not operating tests can be carried out using a video camera inside the chimney shaft. A thermo-visual camera will reveal variations in temperature of the steel chimney shell, which will be highest where there are cracks and lining losses. These places can be located and after the chimney has been taken out of operation visual images can be provided by a video camera.

Apparatus to assess the technical condition of chimney linings invented by a team of engineers from the Polish power station in Belchatow is shown in Figure 6.185. The camera is suspended from the shaft top, on carrying rods appropriate to the shaft diameter. The camera is lowered into the shaft on a scaled rope (Figure 6.186) and the camera rotates 360°. The condition of the lining is displayed on the monitor. Where lining damage is noticed the video camera may be stopped and

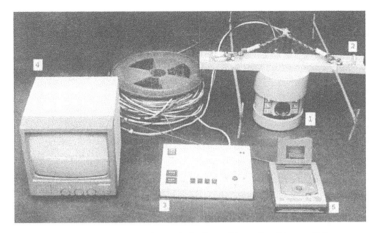

Figure 6.185   Equipment to evaluate the technical condition of a chimney lining:
1 video camera, 2 supporting structure, 3 control unit, 4 monitor, 5 disk station

(a)   (b)

Figure 6.186   Inserting a camera in a chimney: (a) general view, (b) at the chimney top

a picture can be taken (Figure 6.187). Numbers on the images of the chimney lining state the angular displacement of the camera in respect to the starting point.

(a)

(b)

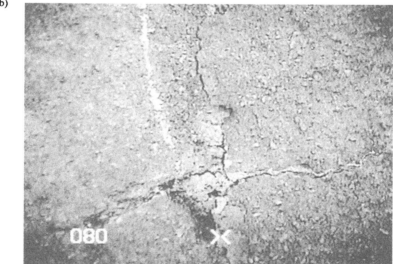

Figure 6.187 The damaged chimney lining: (a) loss in the broken shaft, (b) a crack

6.2.4.3 Repair and reinforcement methods for steel chimneys

Steel chimney structures and operational requirements mean that some general and some particular principles must be considered when planning repairs and designing reinforcements for existing chimneys. Some of these will now be described.

1 Steel shafts susceptible to vibration or showing lateral vibration in strong wind gusts (from vortex excitation) should in principle be equipped with a vortex generator, or a vibration damper, or at least with guy ropes or some other structures to prevent vortex excitation. This applies not only to new, detached chimneys, but also existing chimneys. Vortex generators (turbulence breakers) promote the random breaking away of vortices, so they do not lead to the building of lateral vibrations. Mechanical dampers change the dynamic characteristics of the structure, thus reducing vibration. The dynamic characteristics of the structure may be also altered by changing its weight, rigidity and dynamic pattern. Change of weight can be achieved during repairing to chimneys with extensive corrosion, or in new chimneys during erection.

2 In steel chimneys, vibration is also indicated by fatiguing (cracks), especially at vee-shaped structural and indentations, where local stresses tend to focus. All necessary holes in the chimney shaft should have rounded and ground-out rims, any redundant holes should be closed. All shaft openings need to be framed suitably. Reinforcing elements and welded joints in the shaft should be designed as fillet welds of thickness not more than half of the thinner elements joined together. The welds should not be too close to one another, even if both sides of plate are taken into consideration. The welding process engineering used must ensure the smallest possible level of welding stress. Therefore the order of welds should be in compliance with the process guidelines, and the thinnest possible electrodes and welds, as well short beads, should be used. Weld faces should be ground.

3 Corrosive wear of steel chimneys shortens their operational life by reducing the section of the load-bearing shaft and by degrading the strength qualities of the steel. Evaluation for the purpose of continued operation the partially corroded shafts should be verified by analytic calculation, taking into account current and foreseeable corrosion losses. Changes in geometrical and strength characteristics may occur where there are highest losses in the shaft walls, not in the sections where the largest internal forces occur. Chimneys with corrosive wear that significantly affects the load-bearing capacity and stability of the shaft and segment couplings may be passed for further operation only temporarily and with reservations. Corrosive damage to shaft plates and segment couplings cannot be repaired.

The radical solution, often economically justified, is to dismantle the damaged segments and fit new ones (of proper construction, made of proper materials and with suitable couplings). A temporary solution allowing the damaged chimney to operate until the design has been completed include:

- Shortening the shaft via the removal of the upper segments or the segments with the walkway
- Altering of static system, such as by additional bracing or support, etc.
- Locally reinforcing damaged elements, such as by welding on additional elements (ribs, pads, clamping rings etc.).

4 The rate at which the steel shaft is damaged by the corrosive environment (from the presence of sulphur compounds) inside the chimney may be reduced by ensuring that the temperature on the internal chimney surface remains above condensation point. This may be achieved by proper insulation or by warming the shaft from the inside via segment couplings. Thermal insulation via an airtight outer shell made of the thin steel plate or aluminium plates is also possible. However, a shaft insulated in such a way may prove effective only for a limited number of chimneys, because this solution requires high temperature, high velocity of exhaust gases flow and few operational stoppages.

- Steel shafts are very particular structures due to the nature of their construction, so repair methods should use their light-weight, limited height and ease of dismantling via commonly available cranes. In certain cases, it is possible to dismantle the most damaged parts or even the whole shaft, carry out the repair and reinforcement work on the ground, and reassemble and re-erect the chimney.

6 Each repair or a replacement of an existing chimney should be proceeded by an economic analysis that takes into account not only the cost of repairs, reinforcement, protection and maintenance works needed for future operation, but also the cost of operational stoppages and other interruptions.

6.2.4.4 Exemple of repairs to damaged chimneys

Example 1   A detached chimney of height 80 m and shaft diameter 2.91 m, was exposed to winds that produced resonant vibrations. The repair was carried out several years after the commencement of its operational life. It involved mounting a spiral vortex generator on the shaft, consisting of flat bars $300 \times 300 \times 10$ mm welded along the screw line of 5D wavelength (Figure 6.188(a)). The wings of the vortex generator on the shaft are shown in Figure 6.188(b).

Example 2   A chimney of height 70 m, supported by a tripod, showed excessive deflection and vibration due to faulty collar couplings of the shaft segments. Deformed collars caused gaps in the circumference from 4 to 13 mm (Figure 6.189(a)). Temporary reinforcement and stiffening of the field collar was carried out by fitting an asbestos rope into the gaps and welding flat bars to the collars (Figure 6.189(b)).

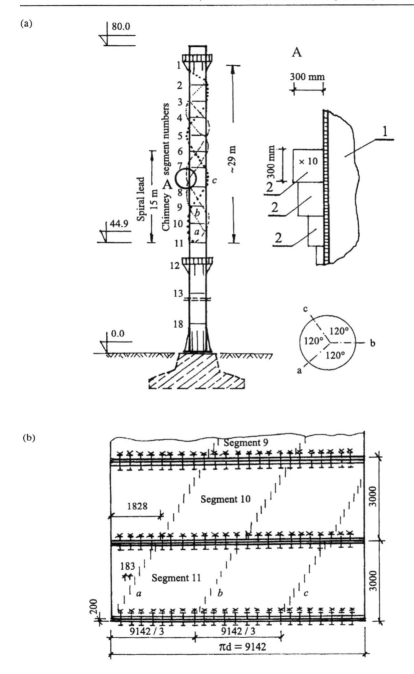

Figure 6.188   Chimney with vortex generators in its part: 1 chimney shaft pipe, 2 vortex generator's rings (a) main view and detail A, (b) location of wings on the vortex generator

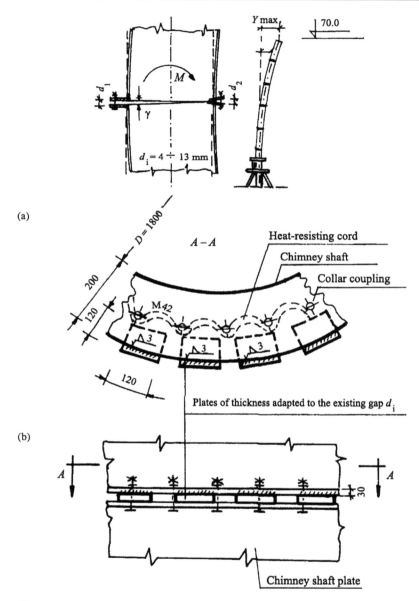

$Y max_t$ 70.0

d_1

M

γ

d_2

$d_i = 4 \div 13\ mm$

(a)

$D = 1800$

200

120

$A - A$

Heat-resisting cord

Chimney shaft

Collar coupling

$M\,42$

$\triangle 3$ $\triangle 3$ $\triangle 3$

120

Plates of thickness adapted to the existing gap d_i

(b)

A A

30

Chimney shaft plate

Figure 6.189   Repair of a chimney of height 70 m (a) schema for a chimney segment damaged flange couplings, (b) the temporary repair

Example 3   The upper part of a chimney shaft above the support tower was broken. This damage was caused by faulty in the design and execution of welded joints between the shaft pipe and the collar (Figure 6.190(a)). The damaged shaft was reconstructed as shown in Figure 6.190(b).

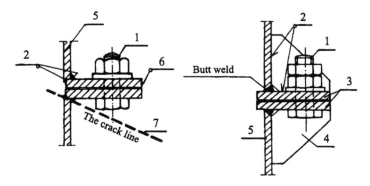

Figure 6.190   Detail of the chimney segments flange coupling: (a) before emergency,
(b) after repair: 1 screw, 2 fillet weld, 3 flange, 4 ribs, 5 chimney shaft pipe,
6 weld executed at construction, 7 point of fracture

6.2.5   Assessment and refurbishing of steel masts

6.2.5.1   Periodic testing of the technical condition of masts and characteristic repair works

Testing of the technical condition of a mast, based on the outcome of a visual inspection, should be carried out once a year. The inspection should take place in the early spring and be preceded by surveying the verticality of the mast shaft from at least two orthogonal directions, and measurements of the initial tightness of the guys. It is advisable that visual inspections be carried out after each occurrence of strong winds and at the end of periods of thaw. Measurements of wall element thickness are recommended every five years. The scope of tests and testing intervals are only approximate and should be scheduled for each mast separately by a structural engineer or the qualified person responsible for the care of the mast.

During the technical survey of the mast shaft, the following should be taken into consideration, as they can influence decisions about repair:

- Corrosion losses that diminish the cross-sectional area of an element or damage to the anti-corrosive coatings caused by the ice peeling, both of which present hazards to the carrying capacity of structural elements (Figure 6.191)
- Local deformation of elements that increase the eccentricity of loadings
- Weld cracks
- Delaminations and cracks in mast elements or their walls
- Damages to legs, welds and bolts, and local deformation of structural elements to which the guys are fixed
- Damage or loosening of screws
- Clearances or gaps in butt and overlapped joints
- Ovalization of screw or rivet holes

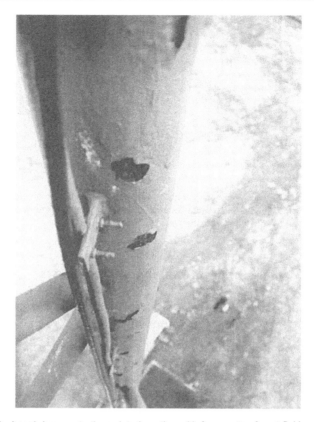

Figure 6.191 Local damage to the painted coating with fragments of coat flaking with icing

- Damage to fittings, antenna systems and fastenings, ladders, railings, platforms, barriers etc.

When the foundations and anchorage are inspected the following should be taken into consideration:

- Horizontal displacements and foundation settlement
- Anchorage corrosion – occurs mainly at places in contact with concrete
- Corrosion of tightening equipment
- Anchorage deformation
- Cracks and fissures in the concrete
- The condition of the concrete surface
- The condition of the foundation insulation, chiefly in the area of concrete and ground contact
- The condition of tightening equipment threads
- The condition of the main insulation or the shaft footing hinge
- Condition of the shaft base anchors.

The guys may be judged to require replacement or repair in cases of:

- Local corrosion that reduces the cross-sectional area of the guy stays. This mostly occurs at flared ends and places where corrosion-preventive grease is missing. The growth of bushes at the base alterations of the guys, may increase losses of protective grease due to branches rubbing the guys (Figure 6.192), so such bushes should be cut down:
- Surface corrosion and damage to external zinc coatings
- Corrosion or damages to bolts, rope thimbles, splicing shackles, clamps etc.
- Break of continuity in more than 5 per cent of strands in one or more bundles over a length of 30 diameters
- Fracture of a guy rope strand
- Local rope unlaying
- Mechanical damage such as break or bump
- A guy rope that has slipped from its pipe flare or coupling

Figure 6.192   Bushes growing near the guy anchors damaging the anti-corrosion protection on the guy ropes

- Negative result of measurement of inner continuity performed using non-destructive methods
- Breaking of an insulator or deformation of an insulator housing
- Abnormal behaviour of guys when tightened, abrupt changes of strength disproportional elongations, rope twists, etc.

It is generally accepted that the mast guys, the elements mostly endangered by fatigue, should be replaced every 20 to 25 years irrespective of their actual technical condition.

The next step after the decision to repair the mast has been taken should be a thorough technical assessment, preferably with an evaluation of a number of repair methods. The repair should be carried out on in accord with a detailed technical and work organization plan.

The repairs commonly carried out on the masts include:

- Upgrading works connected with service changes
- Replacement of aerial systems
- Mounting of new antennae or service platforms
- Guy adjustment and replacement
- Exchange of insulation
- Corrosion protection work
- Reconstruction of mast superstructure
- Repair or replacement of shaft elements.

Such work is not always limited to the checking the structure, but should also consider the forces resulting from new loadings and the mounting of additional supporting structures. In many cases the mast has to be 'formed' again, i.e. the guys have to be re-established and mast displacements calculated. Where the shaft has additional vertical loading and it remains possible to increase horizontal displacements, the initial tightening of the guys may be reduced. If new loadings exceed the admissible strains, the initial draw of the guys may be increased, if there is some available reserve in the shaft carrying capacity. Occasionally, a more interacted solution is required: the draws of some horizontal guys are increased, some are loosened, and others are left unchanged. Increasing the initial tightening may require increasing the rope thickness, changing bolts or anchors and the loading on the foundation guys.

Mast repairs or upgrading require the use of either special equipment or machines rarely used on other steel structures: crawler cranes, transport trolleys moving on guys, jibs to mount guys at close distances, bolt extrusion devices, dynamometers, rope and roller systems, two-way hydraulic motor operators. Correct design and use of such equipment requires thorough knowledge of mechanical engineering, machinery construction and crane components.

In repairing masts little scaffolding is ued, and then it must be individually designed. Most work is carried out directly on the structure using mountaineering methods (Figure 6.193). Due to this, the safety measures are very important. All

Figure 6.193   Repair work on the mast: no-scaffolding is being used (from a catalogue of
the Ramball company)

workers must be equipped with special suspenders, with the bridle on the chest
and safety belts to which the auxiliary equipment used can be attained, and which
can be used as additional fastening in special situations. If masts have unprotected
ladders, safe ascent systems with running guides and ropes with self-wedging or
inertial safely blocks should be used.

Many works are carried out while the masts remain in operation. This requires
that all safety regulations for such conditions be observed. Electromagnetic field
intensities to which it is permissible to expose workers depend on the exposure

time. To extend this time, the power of the emitting equipment should be reduced, and the quality of emission should so be lowered. This needs to be agreed with the manager of the mast and organized in such a way as not to introduce any accidental changes. Field intensities and exposure times are subject to various regulations.

6.2.5.2 Testing and refurbishing mast guys

Work on guys usually involves repairs and maintenance [98]. The initial tightening is checked and adjusted, ropes are tested, measurement of rope thickness carried out, insulation replaced, and corrosion protection layers made good. The guys that failed testing or have exceeded their life are replaced.

To carry out visual inspections and some maintenance works, special trolleys that move on the guys are mounted. This requires that the mast structure to be checked for the loading transferred by the trolley ropes. A trolley mounted directly onto the guy rope would destroy the corrosion protection coating, so it should be mounted on additional rope running from the top of the mast. A schema for such trolley is shown in Figure 6.194. During the repair works, in order to decrease the vertical loading of the shaft and the tension in the ropes, the guy ropes are slackened by 10 to 20 per cent in all directions. Adjustment of the verticality of the shaft is easier and guy ropes are more easier to operate. The mast is then said to be in soft state. However, the guy ropes must not be slackened too much, so as not to cause the loss of mast shaft stability. The limiting case must be determined computationally in specifying the techniques to be used in the repair.

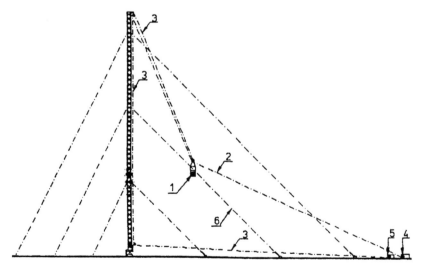

Figure 6.194 Trolley to carry out maintenance and test on tightening ropes: 1 trolley, 2 lead rope, 3 traversing gear cord, 4 lead rope hoisting winch, 5 traversing gear hoisting winch, 6 serviced guy

Usually in repairing guys, mounting guys are fitted to take over the function of the previous ones during the overhaul works. Introducing such a stay rope to the structure requires gradual slackening of the existing guy and simultaneous tightening of the mounting guy, so as not to alter the dynamic system of the mast shaft. Even more difficult is the reassembly, and final tightening because the ropes must not remain in contact with one another, in order not to damage their corrosion protection coatings. For the same reasons contact between guy ropes and antennae or platforms protruding beyond the shaft profile must be avoided. Solutions to this problem may entail to slight changes to the angles in the horizontal plane, using special cantilevers mounted on the mast shaft, retaining the mounting guy, or reconstructing damaged surfaces after completing the mounting work.

The mounting guy is fastened simultaneously to two curbs separate from the curbs of the guy rope under repair (Figure 6.195) or to the same curb, but on a different level. If slackening of the guy ropes is not sufficient local strengthening of the shaft is required. If this still proves insufficient, additional beams can be used to transfer forces from the mounting guy to the mast shaft. The other part of the mounting guy is usually anchored to the same foundation to which the first guy was anchored. If this is impossible, one or more ground anchors should be fixed nearby.

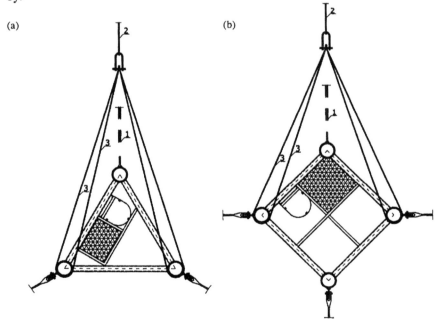

Figure 6.195 Examples of fitting a mounting guy to the chimney shaft: (a) a mast of triangular cross-section, (b) a mast of square cross-section, 1 replaced guy, 2 mounting guy, 3 mounting rope shring around the curb rod of the mast shaft and joined to the mounting guy by a shackle

The carrying strength of the anchors should be proved experimentally before any ropes are attached to them, by applying a test load. The mounting guys and final guys, insulators, and guys between the insulators should also be test loaded before fitting them to the mast. Moreover, all equipment used the during repair, such as rope blocks, dynamometers, servo-motors, extension arms, shackles, bolts and connectors, etc. should be tested before any work is begun. To shorten the preparation time, the carrying-strength tests should be carried out jointly for all elements, as a unit. To test guy ropes, either the existing foundations or the new ground anchors can be used. When lower strength is required to carry out tests, caterpillar tractors can be used.

The test load shall be at least by 25 per cent greater than the highest design load. With the final guy ropes 'overstraining' as distinct from the load test, needs to be carried out to reduce relaxation in the ropes. In this operation the applied load is increased up to 85 per cent of the nominal force, fully stretching the rope. After overstraining the guy ropes should not be reeled onto the drum because this slackens the wires in the rope lay and thus partially lessens the acquired stretching. This is why the overstraining stands are located close to the mounting point. From overstraining, a rope will be lengthened by 1 to 2 mm/m, which means that guy ropes will need to be cut, their ends fastened, and their carrying capacity verified again

(a)

(b)

(c)

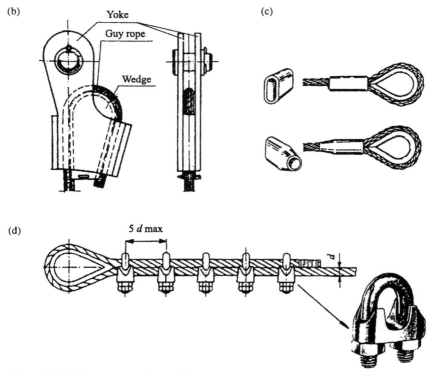

(d)

Figure 6.196   Commonly used guy fitting

to meet the design requirements. Sometimes due to the lack of proper stands, the overstraining test is not carried out. In such cases during the first operational period the guy ropes will need to be tested more often and the design of tightening equipment must take into account a reserve to allow for the elongation of ropes.

The guy rope ends formed into flare (cone- or pear-shaped) couplings (Figure 6.196(a)), self-blocking terminals (Figure 6.196(b)), loops with thimbles fastened on a special press by aluminium coupling (Figure 6.196(c)), or loops fastened by means of bow clamps (Figure 6.196(d)). The connections shown in Figures 6.196(b) or 6.196(c) may be used as hybrids by combining them with the connection in Figure 6.196(d).

For the final guy ropes of high masts flare couplings are used only if made from poured liquid metal alloys (Figure 6.197), chiefly zinc, tin, aluminium, lead and antimony alloys. Some companies also use plastics to finish the ends. The guy rope in the sleeve exit should be secured by a mild wire band, while beyond the band the wires should be unlayed. The end of each wire should be bent, covered with salammoniac, immersed in hot zinc and then inserted in the coupling flare and the metal alloy poured over it. Before pouring with hot alloy, the coupling

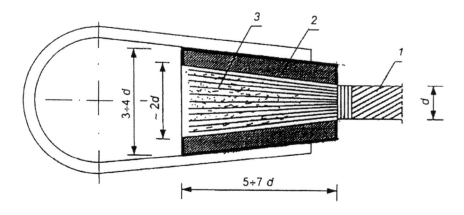

Figure 6.197 Flaxed coupling of the guy (poured liquid metal alloys), main dimensions:
1 guy rope, 2 metal sleeve, 3 metal alloys

should be heated up to a temperature of $150 \div 200°$C. With alloy-poured couplings, single-wire ropes with steel cores are best – other cores should be removed.

Guy ends formed into loops with thimble clamps are used for masts of lower height and for mounting guys. For such couplings – hard rope, i.e. hard single-wire ropes, with a steel core, or a rope with wires thicker than 4 mm should be used. Such ropes are permanently strained, or even damaged, when the loops are formed because thimbles and wedges of bending diameter ($D = 4 \div 5d$) (d = rope nominal diameter) are appropriate only for mild many-core ropes with textile cores of thin wires. Therefore thimbles for rope rolls of diameter $D > 12d$ should be used instead. Rope clamps for hard-core ropes do not produce a proper rope deflection under the bow, so the loop is held by frictional forces. Therefore such ropes it is necessary to check the clamp tightening strength using a torque spanner for example. Mild rope cores are impregnated with grease, which will be saturated when the clamps are fastened so that frictional forces will be reduced. For this reason in the vicinity of the clamp fastening the ropes should be saturated and washed with a special solvent. In the case of immovable clamps, they should additionally be painted with coatings suitable for galvanized surfaces, which apart from providing corrosion protection also indicate whether the rope is sliding out of the clamps. Even when the clamps have been correctly fastened, such ropes tend to show a reduction in tightening strength after some or more loading cycles, due to the shifting of the component wires under the influence of the tensional forces. The clamps have therefore need to be regularly checked for tightness, especially at the mounting guys.

For each loop at least five clamps should be fastened every $5d$, and the clamp bow must be fastened on the shorter end of a rope. The first clamp should be placed as close as possible to the thimble.

All guy ropes shall be equipped with tightening devices used for mast rectification and guy adjustment during operation. These devices usually serve as guy anchors, and when properly used may provide measurements of guy strength. There

(a)

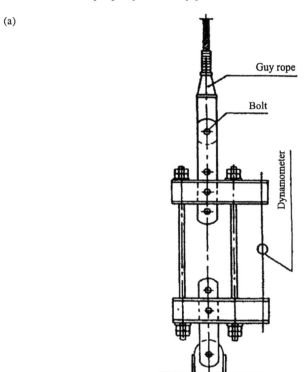

Guy rope

Bolt

Dynamometer

(b)

Figure 6.198 Guy tightening unit with a dynamometer fitted between the brackets (outside the guy ropes): (a) schema, (b) view

are several ways of measuring the forces on guy ropes. The most frequently used, and at the same time the most accurate, is the method of direct measurement. A dynamometer is fixed to the tightening device (Figure 6.198), between the guy and the foundation (Figure 6.199). The advantage of the solution shown is that it decreases lateral vibration by using a rod made of two rectangular pipes that fit together telescopically (Figure 6.200). This rod was mounted after the guy ropes have been tightened, and replaced a dismantled servo-motor.

(a)

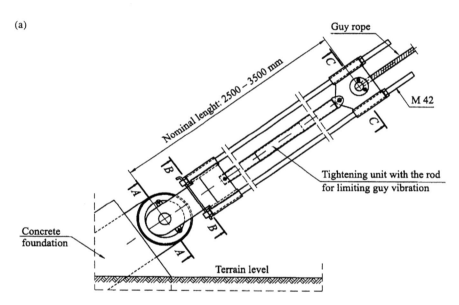

(b)

(c)

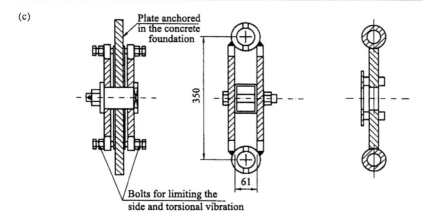

Plate anchored in the concrete foundation

350

61

Bolts for limiting the side and torsional vibration

Figure 6.199   Guy tightening unit with a dynamometer fitted between the guy and foundation: (a) construction, (b) main view without the stabilizing rod, (c) mounting guy fitting eye

Figure 6.200   Tightening unit of Figure 6.199 with a rod limiting vibrations of the guy

Guy rope strength may be also determined by measuring its deflection at any point f, angle of inclination α, the distance from the lower point of fastening to the point of the measured deflection d and the projection of the chord in the horizontal plane L.

$$S = \sqrt{(\frac{qL}{2\cos\alpha} + H\sin\alpha)^2 + (H\cos\alpha)^2} \qquad (6.1)$$

where

$$H = \frac{q(Ld - d^2)}{2f\cos^2\alpha} \qquad (6.2)$$

The force S in the guy is given by equation (6.1), where q is the specific weight of the guy [kN/m].

This method is labour-consuming and not very precise.The force can also be measured by dynamic methods. It is theoretically possible that the magnitude of forces can be calculated via the free vibrations of the guy. However, in practice this is too difficult for high masts so another dynamic method is used. This method makes use of the phenomenon of the of shock wave propagation in an elastic medium. A shock wave in a guy rope is generated by a strong gust fitting the rope near its lower fitting and the time for it to pass is measured by a stop watch. The force calculated in the guy on the basis of this phenomenon is given by equation (6.3):

$$S = \frac{q \cdot n \cdot s^2}{g \cdot t^2} \qquad (6.3)$$

where
- s Total length of the guy
- n Number of oscillations of the shock waves in the guy rope
- t Time taken for n oscillations of the shock wave to pass
- g Gravitational acceleration
- q Specific weight of the guy.

A guy being prepared as a replacement should have additional corrosion protection. For the guy ropes, only galvanized wires should be used. Such protection though good for steel structures, is not safe for guy ropes, which may rub against another. Because guy ropes exhibit deformability, they require an elastic corrosion coating that adheres well to the surface and penetrate to the strands inside. Pastes and greases are therefore good anticorrosive substances. The grease isolates the metal from the environment, and the presence of a corrosion inhibitor will slow the pace of corrosion. The grease also protects the strand inside from dust and sand. Because, friction between strands is diminished, the whole rope becomes elastic, and the internal stresses, as well as the wear on the friction surfaces, are reduced. The greases should be applied under pressure, using a special head on the rope. However, in practice the ropes are often drawn through tubes of the grease. These maintenance works are usually done outdoors, just before the ropes are drawn onto the mast the coiling ropes is not advisable because as with the coiling of ropes after overstraining, this can lead to slackening.

The replacement of guy ropes should always be carried out working from the bottom of the mast to the top. It is not important whether the guy ropes are exchanged one level at a time, or all guy ropes are in one direction but on different levels. It is essential that the forces on the guys are controlled on the level currently being replaced. In general they force measurements should be made on all guy ropes from a given level, but the fittings to replace the guy ropes can be located in one place, common for all guy ropes of a given direction. Such a solution is preferred by the contractors, although it requires an additional rectification of the mast after the operation has been completed.

A basic rule for the replacement of guys is that measurement of forces and surveying the verticality of the mast shaft be done simultaneously. Any deviations of the shaft from the plumb-line should not exceed 1/1500 of the shaft height and 1/1750 of the distance between the guys, at a wind speed of not more than 3m/s. It is obvious that when wind loadings occur the deviations are much larger and may range up to 1/100 of the shaft height or the distance between the guys. Slackening or tightening the guys is carried out in 5 – 10 kN steps, and after each step the shaft deviation should be measured in two orthogonal directions. The sequence of works in the replacement of guys with the mounting guy ropes is as follows:

- Attach the mounting guy according to the method shown in Figure 6.195
- Slacken the old guy rope and tighten the mounting guy by stages, until the mounting guy rope takes over the tensional forces
- Dismantle the old guy and hold the mast with the mounting guy
- Mount the new guy rope
- Tighten the new guy and slacken the mounting guy by stages, until the new guy over the tensional forces,
- Dismantle the mounting guy
- Correct the tightening and repair the anti-corrosion coating.

In some cases it is possible to replace the old guys directly without using the mounting guys. To do this, the guys should be fastened to the mast as shown in Figure 6.201(b). With mast that are not already adapted to such replacement, this operation is slightly more laborious because at first the new guy plays the role of the mounting guy, while next this role is played by the old one. The sequence of works in using this method is as follows:

- Indirectly fasten the new guy to the mast shaft, as shown in Figure 6.195
- Gradually slacken the old guy rope and tighten the new one

(a)   (b)

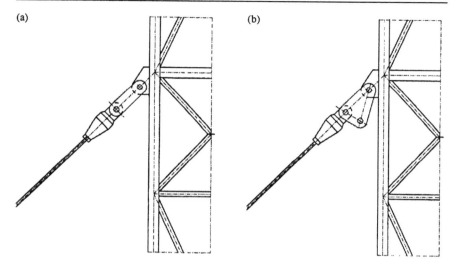

Figure 6.201   Guy connected to the mast shaft: (a) previously used method, (b) new method which makes replacing the guy easier

- Indirectly attach the old guy to the shaft, as shown in Figure 6.195, and release it from the shaft lifting eye
- Gradually tighten the old guy rope on the indirect joint, and slacken the new guy rope till it is quite loose
- Attach the new guy rope to the mast lifting eye and remove the intermediate fastening
- Gradually tighten the new guy and simultaneously slacken the old one until the old guy hangs loosely
- Dismantle the old guy and if necessary rectify the new guy
- Repair the anti-corrosion coating at places where clamps were fastened.

The consideration is the replacement of the guys in the so-called collision directions, which arises when the new mast is dismantled or erected near the existing one. The guy ropes cross in the horizontal plane and interleave in the vertical plane. To interleave the guy ropes, additional transport ropes are normally used to connect the shafts of both masts, fitted with the declination towards the guy rope requiring replacement. The guy fastened to this rope, by shackles or a special trolley, will cover down under its own weight, protected only by a braking line. If the point of crossing of the guys is near the shaft of one of the masts, special cantilevers are used, which extend forward from the shaft to beyond the collision point.

It may happen that, even at low wind velocities, the guys will start to vibrate. To guard against such vibration either a furling line or bobs are used. The works are carried out from a transportation, trolley basket (Figure 6.194) or a mounting trolley basket suspended from a crane hook. Travelling cranes with telescope exten-

Figure 6.202   Fittings to replace the insulator on the guy (photo: K. Dobiszewski)

sion arms are capable of raising a basket to a height of 120 m, but they require a hard-surface access road and are rather expensive. If hanging bobs are used, either double suspension or the suspension of protective nets is advisable. Furling lines need to be tightened such that they do not distort the geometry of the guy ropes. Replacement of isolators mounted on the guys requires the use of special beams (Figure 6.202).

6.2.5.3 Repairs and refurbishing of mast shafts

The majority of existing masts were designed according to regulations that are no longer in force. In such cases, issuing a technical opinion of new operating conditions requires complete calculations for the whole mast [98]. The replacement of an antenna system or mounting an additional parabolic antenna does not significantly increase the structure loading. What is more dangerous is mounting many TV antennae on the mast top, or mounting many antenna systems over sections several metre 9 long. To avoid problems in such cases, early co-operation below the structural engineer the mast operator and the construction engineer is necessary, because there are many seemingly equivalent antenna systems that from the point of view of the structural engineer are quite different. Most masts have some reserve carrying capacity, but the permissible deformation may have been already used or even exceeded. Though new antenna systems may be lighter, their wind pressure surface area may be larger. Therefore the initial guy rope tightening needs to be increased, which means greater loading on the shaft.

There are many ways to strengthen a shaft: strengthen the most-loaded struts decrease the unsuitable bending length of struts by increasing the number of lattices, mounting on additional level of guy ropes and replacing the old ones. The principles governing the strength of struts rods under loading can be found in a number of publications [99], [100] and are applicable to mast constructions. It is very important to maximally unload the strut that is to be strengthened by choosing favourable weather conditions in which to carry out the work, or by introducing forces that work in the opposite direction and at the same time monitoring all factors that influence the local stress (velocity of wind, while forces on the guy ropes, pressure in the unloading devices, etc.).

Unloading excessively loaded struts is not very effective and can involve strengthening of an unacceptable size. In such a case better results are achieved when the strut is replaced completely, although such an operation is much more complicated and dangerous. It requires devising a structure that takes over the load after the strut has been taken out, and gradually transferring the loads from one structure to the other. Shaft strengthening by mounting an additional level of guy ropes are very case-specific, differing for each mast, so it is difficult to set out general rules and procedures. Such strengthening is not always effictive, and occasionally may increase the vertical load on the shaft, so this method gives better results when use to reduce deformation rather than to reduce loads.

Sometimes the height of a mast has to be increased. The increase in height may result from placing on antenna on the top of the shaft or from fitting additional segments. The same methods, may be used as in mounting the mast, notably using creeping crane of sufficient height. It may happen when the antennas are to be mounted, that the cranes lift capacity is too small or it cannot creep the surface. In such cases, an additional structure can be mounted on the mast top in a form of a narrow gantry, which is then raised to the required height by a rope and pulley system and subsequently fixed by screws and bolts. On the top of this structure the lead rope and pulley blocks of the lifting mechanism are mounted. The antenna is lifted as shown in Figure 6.194 for a trolley basket. Another method is to increase the height of two mast walls of a mast triangular section or three walls of a mast of square section. The crane creeps onto them these cracks and through the empty wall the antenna is placed inside. High-tensile bearing bolts widely used to build up the superstructure of a mast from small, manually or semi-mechanically elements.

The replacement of a main insulator involves raising the shaft by a couple of centimetres. This introduces additional tensional stress in the guy ropes which consequently apply additional compression to the mast shaft. To avoid the harmful effect of this, the guy ropes have to be adequately slackened before starting the work of replacing the main insulator. Hydraulic lifting jacks of sufficient lifting power should be inserted under the mast curb rods, and on the base the guide poles bracketing the lower parts of the shaft should be mounted. Each mast should

Figure 6.203   Lower part of the mast shaft adapted to mount a lifting jack used to replace the main insulator

have lower section of the shaft adjusted to enable these procedures to be carried out (Figure 6.203). The lifting jacks used should have identical technical parameters and a common supply from the hydraulic oil pump, but it must also be possible to switch to independent operation in order to carry out upgrading. The jacks must have blocks on their cylinders, such a protective screw to deal with failure of the hydraulic system.

6.2.5.4 The anchor cables of a television transmitter station

The Kráľova Hola anchored aerial mast is located in the Lower Tatras Mountains, Slovakia at a height of 1636 m above sea level. On the basis of hydrometeorological observations, the proposal for cable replacement had to take into account wind speeds of 200 kph and 165 mm frost which considerably exceeds the values normally covered by the valid technical standards.

The mast is of a mono-shaft structure and is anchored on two levels and in three directions (see Figure 6.204). The inner diameter of the shaft is 2800 mm, with wall thickness ranging from 10 to 16 mm. From a spot height +111.95 m, the mast

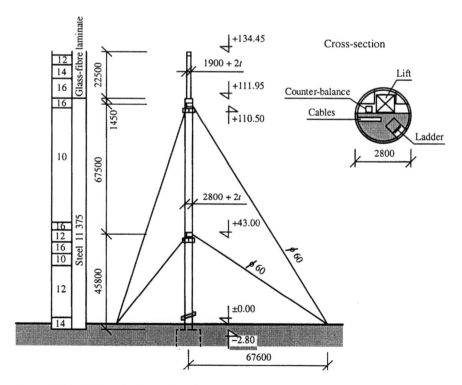

Figure 6.204 Schema for the anchored mast

body is made of a laminate sett. The anchorage of the cables in the anchor blocks uses a rectifying device, which allows changes to the prestressing force in the cables.

In replacing the original cables (Figure 6.205), the various constructional elements of the six anchor cables were modified at both ends for connection to the mast structure.

To the construction site of ŽDB A. S. Bohumín delivered a 293 m long steel single-wound cable with a diameter of 60 mm. The cable core has a closed structure consisting of 90 4.3 mm diameter +1 4.6 mm diameter strands and is covered with a layer of 33 pieces of Z 6 section wires. The cable wires were zinc-coated. According to the attached metallurgical certificate, the nominal strength of the circular wires was 1370 MPa and that of the Z 6 wires 1270 MPa.

Figure 6.205   Cable element with a cable ending

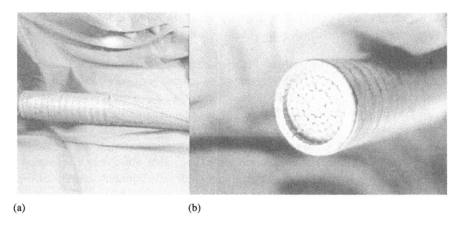

(a)   (b)

Figure 6.206   Defective cable (a) loosening of the surface layer, (b) the core of the cable

It should be noted that in fabrication the manufacturer added one piece of Z 6 wire and 4.6 mm long inner tube of 1 mm diameter to the normal cross-section of a cable with diameter of 60 mm compared to the structure of ON 73 1580 cable. During fabrication of the cable elements, there was a lateral deflection or loosening of Z section wires that form the surface layer of the closed cable (Figure 6.206(a)). This defect in steel cable delivered could be demonstrated by drawing the core against the surfacer layer after the cable had been cut (Figure 6.206(b)).

This phenomenon provided evidence of the non-compactness of the entire cross-section of the cable, consisting of the core of round wires and surface layer of Z profiles. The defects in fabricating the cable resulted in reduced capacity of the anchor cables. It was shown by calculation that if 25 per cent of the Z profiles of the surface layer were removed, the anchor cables would meet the strength requirements.

The efficient functioning of the protective layer was reduced by a partial dissplicing of the surface Z layer, which could reduce life of the anchor cables. During the assessment examination dissplicing in the section was found as far as 4 m from the cable endings. The examiner recommended applying additional anticorrosive protection in the form of gummed bands at both ends of all cables by a minimum of 5 m long, and stabilizing the cross-section of the cable where gumming band ended by clamp rings to prevent further dissplicing.

References

[1]   ENV 1991 – 1: 1994, *Eurocode 1 – Basis of Design and Actions on Structures. Part 1: Basis of Design,* CEN Brussels 1994.

[2]   STN 73 1401: *Designing of Steel Structures.* (In Slovakian). SUTN Bratislava 1998.

[3]   Mrázik, A.: *The Theory of Reliability of Steel Structures.* (In Slovakian). VEDA Bratislava 1989.

[4]   Cornell, C.A.: *Structural Safety Specifications Based on Second-Moment Reliability Analysis.* IABSE Symposium on Concepts of Safety of Structures and Methods of Design. London 1969, Vol. 4, Final Report IABSE, Zürich 1969.

[5]   Ditlefsen, O. and Madsen, H.O.: *Structural Reliability Methods.* John Wilay and Sons, 1996.

[6]   Vičan, J. and Slavík, J.: *Reliability Level of Existing Bridge Structures.* Tenth International Scientific Conference Communications on the Edge of the Millenniums, University of Žilina, 1998, pp. 247–250.

[7]   Vičan, J., Slavík, J. and Koteš, P.: *The Influence of Supervision on the Reliability of Existing Bridges.* (in Slovakian) In 19th Czechoslovak Conference 'Steel Structures and Bridges 2000', Štrbské Pleso 2000, pp. 69–74.

[8]   Frangopol, D.M. and Estes, A.C.: *System reliability for Condition Evaluation of Bridges.* IABSE Workshop 'Evaluation of Existing Steel and Composite Bridges', Lausanne 1997, pp. 47–56.

[9]   Vičan, J. and Koteš, P.: *Effect of Material Degradation on the Structural Element Reliability.* Second International Conference 'Quality and Reliability in the Building Industry', Levoča 2001, pp. 527–532.

[10]   STN 73 0038: *Design and Assessment of Structures in Reconstruction.* (In Slovakian). SUTN Bratislava 1993 (according to ČSN 73 0038, UNM Praha 1986).

[11]   ENV 1993 – 1 – 1: 1992, *Eurocode 3: Design of Steel Structures. Part 1 – 1: General Rules and Rules for Buildings.* CEN Brussels 1992.

[12]   ENV 1993 – 2: 1997, *Eurocode 3: Design of Steel Structures. Part 2: Steel Bridges,* CEN Brussels 1997.

[13]   Nowak, A.S. and Grouni, H.N.: *Calibration of the Ontario Highway Bridge Code 1991 edition.* Canadian Journal of Civil Engineering, Vol 21, 1994, pp. 25–35.

[14] Nařízení c.k. ministerstva železnic ze dne 28. srpna 1904 o mostech železnicových, nadželeznicových a mostech silnic příjezdných a železných nebo dřevěných ústrojinách, číslo 97 z r. 1904. Praha 1911.

[15] ČSN 1230: Jednotný mostní řád – Navrhování mostů, 1937.

[16] ČSN 73 6202: Zatížení a statický výpočet mostů. UNM Praha 1953.

[17] ČSN 73 6204: Projektování ocelových mostních konstrukcí. UNM Praha 1953.

[18] ČSN 73 6205: Navrhování ocelových mostních konstrukcí. UNM Praha 1972.

[19] ČSN 73 6205: Navrhovanie oceľových mostných konštrukcií. SUTN Bratislava 1993 (v znení ČSN 73 6205, UNM Praha 1987).

[20] Ferjenčík, P., Schun, J., et al.: Reconstruction and Reinforcement. Selected Chapters from Metal Structures. (In Slovakian).

[21] Augustyn, J. and Sledziewski, E.: Breakdown of Steel Structures. (In Czech). SNTL – Nakladatelství technické literatury. Praha 1988.

[22] Kaminetzky, D.: Design and Construction Failures. Lessons from Forensic Investigation. McGraw-Hill, Inc. 1991.

[23] Agócs, Z.: Assessment of the Causes of Breakdown of the Support of Steel Structure of Storage Tanks for Filling Micro-ground Calcite and French Chalk into Raj Wagons in the Talcun-Hnúšťa Plant. (In Slovakian). Faculty of Civil Engineering STU, Bratislava 1981.

[24] Agócs, Z., Chladný, E., Baláž, I. et al.: Technical Assessment of the breakdown of the Left Wing of the Left Lock Chamber. The Assessment of Usability of the Lower Gate of the Right Lock Chamber of the Gabčíkovo Water Work. (In Slovakian). Faculty of Civil Engineering STU, Bratislava 1994.

[25] Kálna, K.: Selection of Additional Materials for Manufacture of Steel Structures – Influence of Different Properties of Welded Joints on structure Reliability against Break. (In Slovakian). Zváranie, Svařování 51, No.1–2/2002, pp. 18–23.

[26] Knotková, D.: Failures and Dewfects of Steel Structures Caused by Atmospheric Corrosion. (In Slovakian). Ocelove konstrukce, Vítkovice, š.p., Ostrava, No.4/1990, pp. 101–106.

[27] STN 73 1401. Design of Steel Structures. (In Slovakian). Slovenský ústav technickej normalizácie, 1999.

[28] Agócs, Z. and Brodniansky, J.: Experts' Judgement of the Collapsed Electric Wiring 2×400 kV VO43/V 496: EBO – Bošáca. ZoD No.04-82-99. (In Slovakian). Faculty of civil Engineering STU, Bratislava, April 1999.

[29] Agócs, Z., Chladný, E. et al.: Harmanec, a.s. Paper Mill – Experts' Judgement of the Technical Condition of the Steel Load-Bearing Structures. Mono-blocks after Fire and Proposal of Repairs. (In Slovakian). Faculty of civil Engineering STU, Bratislava March 1995.

[30] Agócs, Z., Chladný, E. and Ferjenčík, P.: Assessment and Refurbishment of Steel Structures. VN 73 0900 Company's Standard. (In Slovakian). Vítkovice, n.p. Ostrava. Faculty of civil Engineering STU, Bratislava 1992.

[31] Agócs, Z.: Application of photogrametry for Assessment of Steel Structures. International Seminar – Interdisciplinary Applications of Photogrametry. (In Slovakian). Proceedings, Kočovce.

[32] Agócs, Z., Chladný, E. et al.: *The Bridge over the Danube between the Towns of Štúrovo and Ostrihom. Assessment of Marginal Fields of the Bridge Structure.* (In Slovakian). Faculty of Civil Engineering STU, Bratislava 1995.

[33] Dutko, P., Ferjenčík, P., Agócs, Z. and Vu Tan Khiem: *Spatial Activity of an Object with a Steel Load-Bearing Skeleton Stressed by Dynamical Loading.* (In Slovakian). Inžinierske stavby No.5, 1979.

[34] Chladný, E. and Agócs, Z.: *Experimental Assessment of Steel Load Bearing Structure of Boilers K1 and K2 of the Kolubara Heating Plant, Yugoslavia.* (In Slovakian). Faculty of Civil Engineering STU, Bratislava 1985.

[35] Vičan, J. and Bujňák, J.: *Assessment and Calculation of Loadability of the Load-Bearing Structure of the Bridge over the Little Danube on Road II/573 in km 44216.* (In Slovakian). Správa pre Slovenskú správu ciest, Správa a údržba Komárno, Technical University in Žilina, 1998.

[36] Šertler, H., Bujňák, J. and Vičan, J.: *Calculation of Loadability of the Bridge over the River Jizera in Krnsko.* Technical University in Žilina, Department of Building Structures and Bridges, 1980.

[37] Tomica, V., Sokolík, A. and Zemko, Š.: *Maintenance and Reconstruction of Bridges.* (In Slovakian). Alfa Bratislava 1992.

[38] ENV 1993 – 1 – 1. *Eurocode 3: Design of Steel Structures. Part 1: General Rules and Rules for Buildings.* CEN Brussels 1992.

[39] STN 73 1401: *Designing of Steel Structures.* (In Slovakian). ÚNMS Bratislava 1998.

[40] ENV 1994 – 1 – 1, *Eurocode 4; Design of composite Steel and Concrete Structures. Part 2: Composite Bridges.* CEN Brussels 1992.

[41] ENV 1994 – 2, *Design of Composite Steel and concrete structures. Part 2: Composite Bridges.* CEN Brussels 1997.

[42] Traini, G. et al.: *New Design Approach for Through Truss Italian Railway Bridges.* Nineth Nordic Steel Construction Conference, Helsinki, 2001, pp. 497–503.

[43] Tomica, V., Vičan, J., Hric, M. and Dobosz, K.: *Experts' Assessment of Steel Bridge Structure of the Device for Impurities Arrest and Removal in the Hričov Water Work.* VŠDS Žilina 1994.

[44] Reel, R.S., and Agorwal, A.C.: *Reliability-Based Evaluation Section of the Draft Canadian Highway Bridge Design Code.* IABSE Workshop 'Evaluation of Existing Steel and Composite Bridges', Lausanne 1997, pp. 17–24.

[45] SR 5 (S): *Determination of Loading Capacity of Railway Bridges.* Service Manual. (In Czech). Czech Rails, s.r., Prague 1995.

[46] Vičan, J. et al.: *Methods for Calculation of Loading Capacity of Existing Road Bridges. Guidance for Slovak Road Administration,* University of Žilina, Žilina 1999.

[47] STN 73 0038: *Designing and Assessment of Building Structures in Refurbishment of ÚNMS SR.* (In Slovakian). Bratislava 1993.

[48] Agócs, Z. et al.: *Experts' Assessment of the Technical Condition of the Part of Steel Structures of Hall M1 Damaged by Fire.* (In Slovakian). SVŠT Bratislava 1986.

[49] Bujňák, J. and Vičan, J.: *Experts' Assessment of the Technical Condition of the Steel Structure of Hall M1.* (In Slovakian). VŠDS Žilina, 1996.

[50] STN 73 0035: *Loading of Building Structures*. (In Slovakian). ÚNMS SR, Bratislava 1993.

[51] Tesár, A.: *The Roof of the Winter Stadium in Bratislava*. (In Slovakian). Inženýrske stavby 1959

[52] Tesár, A., Agócs, Z., Lapos, J., Oláh, J., Kuzma, J. and Fillo, L.: *Experts' Judgement of the Steel and Reinforced Concrete Load Bearing Structure of Winter Stadium in Bratislava*. (In Slovakian). Faculty of Civil Engineering, Slovak Technical University, Bratislava 1989.

[53] Agócs, Z. and Lapos, J.: *Reconstruction of the Winter Stadium in Bratislava*. Faculty of civil Engineering, Bratislava 1987.

[54] Brodniansky, J. and Recký, J.: *Manual for Assessment and Refurbishment of Agricultural Buildings*. (In Slovakian). INTOP, Bratislava 1990.

[55] Agócs, Z., Brodniansky, J. et al.: *Experts' Judgement and Proposal for modification of the Steel Load-Bearing Structures of Objects No.202, 203, 204/3 and 207, s.p. SH Senica, Bratislava*. First Stage – June 1994, Second Stage – June 1995.

[56] Schun, J., Ferjenčík, P. and Agócs, Z.: *Assessment of Roof Steel Structure of a Cinema*. (In Slovakian). Bratislava 1966.

[57] Dutko, P., Agócs, Z. and Lapos, J.: *Experimental Monitoring of Prestressing High-strength Wires of the Roof of a Customs Area*. (In Slovakian). Bratislava 1967.

[58] Brodniansky, J., Nádaský, P. and Shawkat, S.: *Experts' Judgement and Proposal for General Repair of the Roof Load-Bearing Structures of the Regena, Ltd. Service Halls, Pezinok*. (In Slovakian). Bratislava April 1996.

[59] Agócs, Z., Ferjenčík, P. and Chladný, E.: *Refurbishment of the Civil Engineering Department, Slovak Technical University*. (In Slovakian). Bratislava 1987, pp. 407–21.

[60] Ferjenčík, P., Chaldný, E. and Agócs, Z.: *Experts' Judgement of the Steel Load-Bearing Structure of the Smrečina, š.p. Matches Manufacturing Plant Smrečina, Banská Bystrica*. (In Slovakian). Faculty of Civil Engineering. Slovak Technical University, Bratislava 1986.

[61] Agócs, Z.: *Experts' Assessment of the Technical Condition of a Part of Steel Roof Structure of the Hall in Tesla, Orava Damaged by Fire*. (In Slovakian). Faculty of Civil Engineering, Slovak Technical University, Bratislava 1986.

[62] Agócs, Z. and Lapos, J.: *Experts' s Opinion of the Steel Load-Bearing Structure of Conveyer Bridges at NPK, n.p. Duslo Šaľa*. (In Slovakian). Faculty of Civil Engineering, STU, Bratislava 1978.

[63] Agócs, Z. and Lapos, J.: *The Reconstruction of the Steel Load-Bearing Structure of NPK the Conveyer Bridges Attacked by Corrosion*. (In Slovakian). Technický spravodaj. Oceľové konštrukcie Vítkovice, Ostrava 2/80.

[64] Agócs, Z. et al.: *Solution for the Refurbishment of the NPK Conveyer Bridges and their Supports at Duslo Šaľa*. (In Slovakian). Dom techniky ČSVTS. Bratislava 1985.

[65] Agócs, Z., Lapos, J., Baláž, I. and Brodniansky, J.: *A Study of NPK Conveyer Bridges for Duslo Šaľa*. (In Slovakian). Dom techniky ČSVTS, Bratislava 1986.

[66] Agócs, Z. et al.: *Reconstruction of the NPK Conveyer Bridges at Duslo Šaľa*. (In Slovakian). Inženýrske stavby No.11–1991, pp. 380–383.

[67]   Agócs, Z.: *Expert's Assessment of Reconstruction of the Bridge over the Danube Štúrovo-Esztergom.* (In Slovakian). KkaDK, SvF SVŠT, Bratislava 1990.

[68]   Agócs, Z., Chladný, E., Agócs, Z. ml. a kol.: *A Study of Reconstruction of Road Border Bridge over the Danube Štúrovo-Esztergom (Steel Structure).* (In Slovakian). KkaDK, SVŠT, Bratislava 1990.

[69]   Agócs, Z., et al.: *A Study of the Reconstruction of the Bridge over the Danube at Štúrovo (Steel Structure).* (In Slovakian). STU, Stavebná fakulta, Bratislava 1991.

[70]   Agócs, Z.: *Possibilities for Refurbishment of the Bridge over the Danube at Štúrovo-Esztergom.* (In Slovakian). Inžinierske stavby No.2/1992, pp. 387–392.

[71]   Agócs, Z. and Medelská, V.: *Refurbishment of the Mária-Valéria Bridge over the Danube between the Towns of Štúrovo and Esztergom.* Attachment to the Application for Obtaining the Grant from the PHARE programme. (In Slovakian). STU, Stavebná fakulta, Bratislava 1995.

[72]   Agócs, Z., et al.: *Refurbishment of the Mária-Valéria Bridge over the Danube between the Towns of Štúrovo and a Esztergom. A Detail Static Calculation of Steel Structure of the Bridge.* Order No.04.132.99, Bratislava June 1999.

[73]   Agócs, Z., Chladný, E. and Chladná, M.: *The Bridge over the Danube between the Towns of Štúrovo-Ostrihom, DVP.* Order No.04.194.50. Static Calculation 2 – Outer (Original) Bridge Fields – Load-Bearing Structure. (In Slovakian). KKaDK, STU, Stavebná fakulta, Bratislava May 2000.

[74]   Agócs, Z.: *Reconstruction of the Esztergom-Štúrovo Danube Bridge from 1895.* Proceedings of the IASS – MSU International Symposium. Istanbul May 2000, pp. 205–14.

[75]   Agócs, Z.: *Stavba,* Vol. III, No.12/2000, pp.16–19.

[76]   Agócs, Z.: *Refurbishment of Mária Valéria Bridge over the Danube between the Towns of Štúrovo and Esztergom.* (In Slovakian). Staviteľský almanach 2001. KASI 5, SKSI, pp. 139–46.

[77]   Wright, B.H. and Hasteltex, H.F.: *Soc. of Automotive Engineering,* 1953.

[78]   Ziółko, J. and Supernak, E.: *Naprawa zbiorników stalowych na paliwa płynne uszkodzonych wskutek korozji wżerowej, (Repair of Steel Tanks for Liquid Fuels, Damaged Due to the Pit Corrosion),* Inżynieria i Budownictwo No.7/1996.

[79]   Słoniowski, A.: *Ocena zaawansowania procesów korozyjnych den zbiorników na ropę. Ochrona przed korozją, grudzień 1996 (Progress Assessment of the Corrosion Processes in the Crude Oil Tanks.* Corrosion protection, December 1996).

[80]   Horner, J. and Hinger, R.: *Tank Rehabilitation and tertiary containment.* The Second International Symposium on Above-ground Storage Tanks, 1992, Houston, Texas.

[81]   Dave, B. McNeill and Scott, A. McEachern: *Above-ground Oil Storage Tank Maintenance Program.* First International Pipeline Conference – IPC 96, Calgary Alberta 1996.

[82]   API Standard 653. *Tank Inspection, Repair, Alteration and Reconstruction, January 1991 with amendments from January 1, 1992.*

[83]   Ziółko J.: *Die Instandsetzung durch Unterdruck beschädigter zylindrischer Stahlbehälter.* Der Stahlbáu No.11/1980.

[84]   Ziółko J.: *Zbiorniki metalowe na ciecze i gazy (Steel Tanks for Liquids and Gases).* Arkady, Warszawa 1986.

[85] Supernak, E.: *Oprava Oceľovej nádrže po havárijnom poškodení plášťa.* Inžinerske stavby No.2/1996.

[86] Ziółko, J., Supernak, E., Borek, P. and Jędrzejewski, M.: *Repair of Cylindrical Tanks Damaged by Negative Presure Produced in their Inner Space.* International Conference on Carrying Capacity of Steel Shell Structures, Brno 1997, Proceedings pp. 246–51.

[87] Ziółko, J.: *Formkorrektur an Mänteln zylindrischen Stahlbehälters.* Stahlbau No.5/1993.

[88] Ziółko, J.: *Instandsetzung am verformten Mantel eines zylindrischer Stahlbehälters.* Stahlbau No. 6/1993.

[89] Ziółko, J.: *Modelluntersuchungen der Windeinwirkung uuf Stahlbehälters mit Schwimmdach.* Stahlbau No.11/1978.

[90] Tiratsoo, J.N.H.: *Pipeline Pigging Technology. Second Edition.* Pipes and Pipeline International Scientific Surveys Ltd. Beaconsfield 1991.

[91] Raczyński, M. and Bogotko, W.: *Badania odkształceń rurociągów. Rurociągi. (Pipeline deformations tests. Pipelines).* Zeszyty Problemowe Polskiego Stowarzyszenia Budowniczych Rurociągów No.4/1998.

[92] Karpiński, M. and Skrok, K.: *Zalecenia wyboru sposobu naprawy defektów rurociągów przemysłowych. (Guidelines for the Selection of Damages Repair in the Industrial Pipelines).* Inżyniera i Budownictwo No.12/2000.

[93] Runkiewicz, L. and Midak, J.: *Ocena ubytków korozyjnych I zagrożeń eksploatacyjnych istniejących kominów stalowych (Assessment of Corrosion Damages and Operational Use Hazards of the Existing Steel Chimneys).* Inżynieria i Budownictwo No.12/1997.

[94] Włodarczyk, W.: *Kominy stalowe* (rozdział 14.6) w pracy Łubiński, M. and Żółtkowski, W.: *Konstrukcje metalowe część II.* Arkady Warsaw 1992. (Steel Chimneys chapter (14.6) in the paper of Łubiński, M. and Żółtkowski, W.).

[95] *DIN 4133 Schornsteine aus Stahl 1991.*

[96] *Model Code for Steel Chimneys.* CICIND 1999.

[97] Włodarczyk, W.: *Remonty I wzmocnienia kominów stalowych.* XV Ogólnopolska Konferencja Warsztat pracy projektanta konstrukcji Tom II Naprawy i wzmocnienia konstrukcji stalowych. Ustroń 2000 (*Repairs and Reinforcement of Steel Chimneys* XV National Conference 'Construction Designer's Workplace' Volume II, Repairs and Reinforcements of the Steel Structures, Ustroñ 2000).

[98] Dobiszewski, K.: *Remonty i wzmocnienia masztów stalowych.* XV Ogólnopolska Konferencja Warsztaty pracy projektanta konstrukcji Tom I. Naprawy i wzmocnienia konstrukcji stalowych Ustroń 2000 (*Repairs and Reinforcement of Steel Masts.* XV National Conference 'Construction Designer's Workplace' Volume I, Repairs and Reinforcements of the Steel Structures, Ustroń 2000).

[99] Ziółko, J.: *Utrzymanie i modernizacja konstrukcji stalowych. (Maintenance and Modernisation of Steel Structures).* Arkady, Warszawa 1986.

[100] Bródka, J.: *Przebudowa i utrzymanie konstrukcji stalowych. (Refurbishing and Maintenace of Steel Structures).* Mostostal Projekt S.A. Politechnika Łódzka, Warszawa, Łódź 1995.

Index